KU-777-764

Contents

A BASIC PRINCIPLES
1 Telecommunication systems   1
2 Line system characteristics   2
3 Radio system characteristics   5
4 Switching system principles   7

B SOME FUNDAMENTALS
5 The decibel   10
6 Frequency, waveforms and filters   12
7 Voice frequencies   19
8 Attenuation and noise   21
9 Amplitude modulation   24
10 Frequency modulation   27
11 Pulse modulation   29

C SWITCHING AND SIGNALLING
12 Step-by-step telephone exchanges   31
13 Reed relay and crossbar exchanges   35
14 Stored program control   39
15 Signalling   40
16 PBXs, PABXs, keyphones, Centrex   43
17 Digital exchanges   46

D CABLE, RADIO AND TRANSMISSION
18 Local distribution networks   51
19 Carrier working: groups and supergroups   55
20 Submarine cables   57
21 Optic fibres   60
22 Radio propagation   64
23 Antennae   68
24 Satellites   72
25 Maritime communications   76
26 Mobile radio systems   78

E MAINTENANCE AND OPERATION
27 Centralized maintenance systems   81
28 Telephone tariffs and system viability   83
29 Forecasting future demands   85

F TELEVISION
30 Television principles   90
31 Colour television   93

G   DIGITAL SERVICES
32 Analog versus digital   97
33 Pulse code modulation   101
34 Primary PCM systems and time-division multiplexing   104
35 Data services   109
36 Packet switching   113
37 Digital strategies for national telephone systems   118
38 Local area networks   119
39 Integrated services digital networks   124

H   DIGITAL FUNDAMENTALS
40 Bits, bytes, words   127
41 Integrated circuits   128
42 Logic gates   130
43 Memories and stores   134
44 Magnetic tapes and discs   137
45 Computers and microprocessors   141

I   TODAY AND TOMORROW
46 Electronic mail   144
47 Voice messaging   147
48 Teletext, Viewdata, Prestel   149
49 Cable television   152
50 High-definition television   154
51 Credit cards and smart cards   155
52 Direct broadcasting by satellite   157
53 Travel or communicate?   160

54 Corporate communications   162

Index   167

3 8023 001693512

Telecommunications Primer

SECOND EDITION

Graham Langley MBE, BSc, CEng, FIEE, MBIM

Books are to be returned on or before
the last date below.

LIBREX —

ANDERSONIAN LIBRARY
WITHDRAWN FROM LIBRARY STOCK
UNIVERSITY OF STRATHCLYDE

Pitman

UNIVERSITY OF
STRATHCLYDE LIBRARIES

384 LAN

K5957

PITMAN PUBLISHING
128 Long Acre London WC2E 9AN

© G. A. Langley 1986

First published in Great Britain 1983
Reprinted (with new section 47) 1984
Reprinted 1985
Second edition 1986
Reprinted 1986

JORDANHILL COLLEGE LIBRARY

All rights reserved. No part of this publication may be reproduced,
stored in a retrieval system, or transmitted, in any form or by any
means, electronic, mechanical, photocopying, recording and/or
otherwise without the prior written permission of the publishers. This
book may not be lent, resold, hired out or otherwise disposed of by
way of trade in any form of binding or cover other than that in which it
is published, without the prior consent of the publishers.

ISBN 0 273 02478 7

Printed and bound in Great Britain at The Bath Press, Avon

D
621·38
LAN

Preface

More and more people are beginning to believe that future prosperity will depend tremendously on the extent to which efficient use is made—on a large scale—of telecommunications and microelectronic processors. The computing and telecommunications industries have in recent years begun to move together; new terms have been coined to describe this movement, e.g. Information Technology or IT in Britain, Telematiques in France (anglicised to Telematics in the USA).

A general knowledge of Telematics is nowadays essential for anyone who wishes to be an effective top executive, in any field; corporate efficiency often depends more on wise use of telematics than on any other single factor. It is no longer commercial sense to leave such matters entirely to the technicians and engineers directly concerned.

This book has therefore been prepared with two main objectives.

First, to give new technical recruits to the profession a rapid introduction and background to the various specialist activities and modern developments which will be facing them.

Secondly, to provide professionals in other disciplines, and non-technical managers generally, with easy-to-read but authoritative papers on many of the technically complex matters which are fast becoming important parts of our everyday lives.

The book has been written in a user-friendly way which takes the reader step-by-step up the technological ladder; I have tried to ensure that no concept is mentioned in passing until it has been defined in layman's language.

The preparation of this 2nd Edition has given me an opportunity to make a few minor revisions to the text, eliminating printer's gremlins, and also to incorporate six additional sections which bring the book as up-to-date as it is possible for a book to be in this rapidly changing world.

Acknowledgements

In no way would I claim to be an expert in all the many subjects dealt with in this book, and there seemed little point in pretending to be re-inventing the wheel. A number of the sections given here do therefore make much use, with grateful acknowledgements, of material given in existing textbooks published by Pitman Publishing Ltd. Those textbooks, all of which are recent and in print, are as follows:

Telecommunication Systems by P. H. Smale
Transmission Systems II by D. C. Green
Radio Systems for Technicians by D. C. Green
Digital Techniques and Systems by D. C. Green
Electronics II by D. C. Green
Electronics III by D. C. Green
Microprocessors: Essentials, Components, and Systems by R. Meadows and A. J. Parsons
Digital Electronic Technology by D. C. Green
Electrical and Electronic Principles by N. Morris
Essential Electronics by G. C. Loveday

In particular, several illustrations originally appearing in these books have been used in this book where applicable, and also a number of text passages used either directly or modified.

Although my debt to all these Pitman authors is considerable, I must record a special thank you to Mr P. H. Smale, whose *Telecommunications Systems* provided much basic information for many sections, including almost the whole of the text and the diagrams for the sections here on Television.

Another source of material, often consulted, was Telephony's *Dictionary* of telecommunications terms, published in Chicago in 1982. My grateful thanks go to the Telephony Publishing Corporation.

I should like also to thank colleagues in the British telecommunications industry who have helped by criticising first drafts and who provided germs for ideas used in these pages. Telecommunications these days is very much a team effort; suggestions have come from so many sources that to name all would be impracticable.

Part A Basic Principles

1 Telecommunications Systems

Telecommunications has been most neatly defined as the technology concerned with communicating at a distance.

The first requirement is for the original information energy (such as that of the human voice, or music, or a telegraph signal) to be converted into electrical form to produce an electronic information signal. This is achieved by a suitable transducer, which is a general term given to any device that converts energy from one form to another when required.

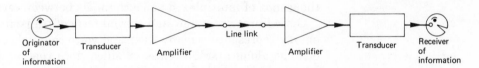

Fig. 1.1 **Basic requirements for a one-way line telecommunication channel**

In a line telecommunication system (fig. 1.1), the electronic signal is passed to the destination by a wire or cable link, with the energy travelling at a speed of up to 60% that of light (depending on the type of line). At the destination, a second transducer converts the electronic signal back into the original energy form. In practical systems, other items may also be required; for example, amplifiers may be needed at appropriate points in the system. Amplifiers do not change the signal from one form of energy to another; they are usually inserted when it is necessary to increase the power level of signals to compensate for losses encountered.

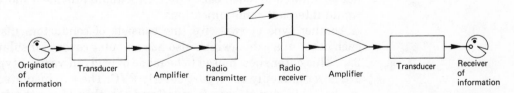

Fig. 1.2 **Basic requirements for a one-way radio telecommunication channel**

For a radio system (fig. 1.2), a transmitter is required at the source to send the signal over the radio link, with the energy travelling at the speed of light, and a receiver is needed at the destination to recover the signal before applying it to the transducer.

At this point it is important to realize that, in both these systems, interference will be generated by electronic noise, and also that distortion of the electronic signal will occur for a number of reasons. These are undesirable effects and must be minimized in the system design.

Simple single-voice-band systems are one-way only (unidirectional), and generally called channels; domestic radio and television broadcasting are familiar examples of such systems.

Other systems, however, such as national telephone systems, must be capable of conveying information in both directions. To do this, the basic requirements must be duplicated in the opposite direction: a pair of complementary channels provide bi-directional communication, generally called a circuit.

When more than one circuit is needed between two points, it is not always economically practicable for another pair of wires or another radio system to be provided. Equipment is available which enables more than one voice channel to be carried on a pair of wires, a coaxial cable, a radio link or an optic fibre. Such multi-channel equipment is called carrier or multiplexing equipment.

Coaxial cable networks, with amplifying stations every few kilometres, now link together most of the cities in developed countries. These networks carry many thousand multiplexed channels.

Radio equipments operating at frequencies much higher than ordinary domestic radio sets and called microwave links are also able to carry thousands of multiplexed voice channels between terminals.

Optic fibres are a new and special form of transmission path in which energy representing many thousand voice channels can travel as pulses of light along a single glass or silica fibre comparable in diameter to a human hair. Optic fibre cable networks are now being installed in many countries, making possible a huge expansion of telecommunications services.

2 Line System Characteristics

The simplest form of two-wire line is produced by using bare conductors suspended on insulators at the top of poles (see fig. 2.1). The wires must not be allowed to touch each other; this would provide a short-circuit and would interrupt communications.

Another type of two-wire line consists of conductors insulated from each other in a cable, which also has an outer cover of insulation (see fig. 2.2). This outer sheath used to be made of lead but various types of plastic are now commonly used, particularly PVC. The two insulated conductors in the cable shown are often twisted together along the length of the cable, and are called a pair.

Many two-wire lines are often wanted between the same two places. These can most conveniently be provided by making a cable with a number of pairs of insulated wires inside it. Sometimes the wires are twisted together in pairs (as illustrated in fig. 2.3) but sometimes they are provided in fours or quads (as shown in fig. 2.4).

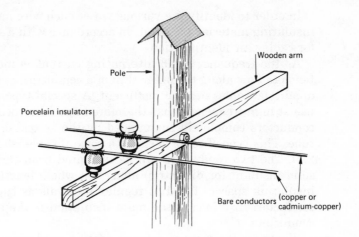

Fig. 2.1 Simple overhead two-wire line

Wooden arm

Pole

Porcelain insulators

Bare conductors (copper or cadmium-copper)

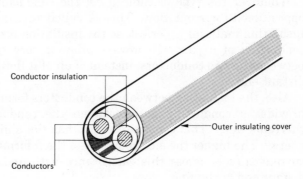

Fig. 2.2 Simple two-wire cable

Conductor insulation

Outer insulating cover

Conductors

Fig. 2.3 Audio-frequency unit-twin cable

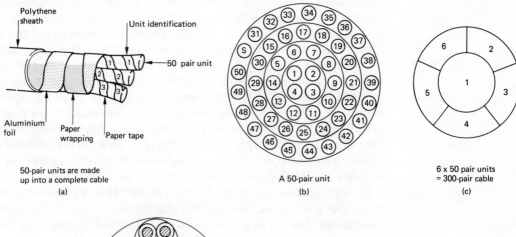

Polythene sheath

Unit identification

50 pair unit

Aluminium foil

Paper wrapping

Paper tape

50-pair units are made up into a complete cable

(a)

A 50-pair unit

(b)

6 × 50 pair units = 300-pair cable

(c)

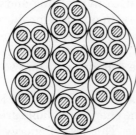

Fig. 2.4 Quad-type cable

In order to identify the various wires, each wire has a colouring on the insulating material around it, in accordance with a standard colour code for cable pair identification.

As the frequency of an alternating current is increased, the current tends to flow along the outer skin of a conductor, and ordinary twin and quad type cables become inefficient. A special type of cable suitable for use at high frequencies has therefore been developed. This has one of its conductors completely surrounded by the second one, in the form of a tube. This type of cable is called a coaxial cable (shown in figs. 2.5, 2.6, 2.7). The two conductors can be insulated from each other either by a solid insulant (or dielectric) along the whole length of the cable or by insulating spacers fitted at regular intervals as supports for the inner conductor. In this case the main insulation is the air between the two conductors.

Whatever the type of cable used, the conductors always have some opposition to current flow. This is called resistance. Furthermore no insulating material is perfect, so the insulation used to separate the two conductors of a pair will always allow a very small current to flow between the two conductors, instead of all of it flowing along them to the distant end.

Also, the insulation between the conductors forms a capacitance which provides a conducting path between the conductors for alternating currrents (a.c.). This capacitance also has the ability to store electrical energy. The higher the a.c. frequency of the information signal, the more current travels across this capacitance path and the less reaches the distant end of the line.

When an electric current flows along a wire, a magnetic "field" is established around the wire. Bringing an ordinary magnetic compass near to a wire shows whether or not the wire is carrying a direct current; if it is, the magnet's needle swings. There is no need to cut the wire and insert a test meter. Whenever the current in a wire changes, either by the switching on-or-off of a one-way or direct current, or the repetitive changes of an alternating current, its accompanying magnetic field is made to change also, and energy is needed for these changes. This is called the inductance of the circuit. If there are other wires nearby, they are affected by such changing magnetic fields; there is inductive coupling between the wires. This is the principle used in transformers, which enable electric power to be transferred from one circuit to another without actual physical contact between the two circuits.

Energy is used up to make the current flow against the resistance along the conductors, and against the insulation resistance between the conductors. Energy is also used in charging and discharging the capacitance between the conductors. In multi-pair cables there is capacitive and inductive coupling between pairs also, so that some energy is passed from one pair to another. These losses further reduce the amount of energy that reaches the end of the original pair and so contribute to the total loss.

In the case of an information signal, all this lost energy has to come from the signal source, so the energy available gradually decreases as the signal travels along the line. This loss of energy along the line is called attenuation. If the line is long and the attenuation is large, the received energy may be too weak to operate the receiving transducer—unless some corrective action is taken.

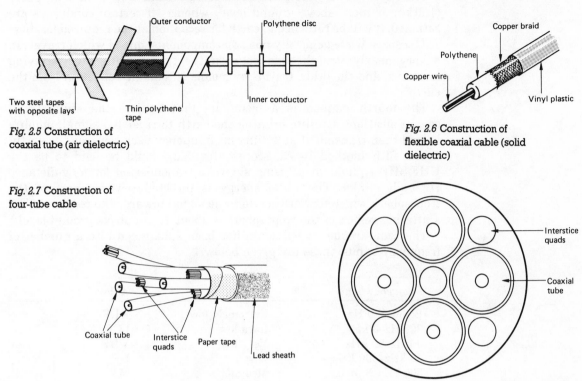

Fig. 2.5 Construction of coaxial tube (air dielectric)

Fig. 2.6 Construction of flexible coaxial cable (solid dielectric)

Fig. 2.7 Construction of four-tube cable

(a)

(b)

3 Radio System Characteristics

When a radio-frequency current flows into a transmitting antenna (aerial), power is radiated in a number of directions in what is called an electro-magnetic wave. This is a complex signal with the same general characteristics as light but of a lower frequency; electro-magnetic radio waves travel at the same speed as light and can be reflected and refracted just as light can be. Some antennae are designed to be highly directional, some are omni-directional. The radiated energy (see fig. 3.1) will reach the receiving station by one or more of five different modes:

1) Surface wave
2) Sky wave
3) Space wave
4) Via a satellite
5) Scatter

The surface wave is supported at its lower edge by the surface of the earth and is able to follow the curvature of the earth as it travels.

The sky wave is directed upwards from the earth into the ionosphere (100 km or more above ground level) whence, if certain conditions are satisfied, it will be returned to earth for reception at the required locality.

The space wave generally has two components, one of which travels in a very nearly straight line between the transmitting and receiving locations, and the other travels by means of a single reflection from the earth.

The fourth method is a technique that utilizes the ability of a communications satellite orbiting the earth to receive a signal, amplify it, and then transmit it at a different frequency back towards the earth.

The fifth method listed, scatter (fig. 3.2), could be said to be the UHF/SHF equivalent of using skywave transmission for long-distance HF radio links. The radio energy is directed towards part of the troposphere which forward-scatters the signal towards the receiver. (The scattering region of the troposphere is about 10 km above ground level.)

The radio-frequency spectrum has been subdivided into a number of frequency bands; these are given below.

Frequency band	Classification	Abbreviation
Below 300 Hz	Extremely low	ELF
300 Hz–3 kHz	Infra low	ILF
3 kHz–30 kHz	Very low	VLF
30 kHz–300 kHz	Low	LF
300 kHz–3 MHz	Medium	MF
3 MHz–30 MHz	High	HF
30 MHz–300 MHz	Very high	VHF
300 MHz–3 GHz	Ultra high	UHF
3 GHz–30 GHz	Super high	SHF
30 GHz–300 GHz	Extremely high	EHF
300 GHz–3000 GHz	Tremendously high	THF

[See Section 6 for *Frequency*]

The surface wave is used for world-wide communications in the low-frequency bands and for broadcasting in the MF band.

The sky wave is used for HF radio communications systems, including long-distance radio-telephony and sound broadcasting.

The space wave is used for sound and TV broadcasting, for multi-channel telephony systems, and for various mobile systems, operating in the VHF, UHF, SHF and higher bands.

Communication satellites are used to carry multi-channel telephony systems, television signals, and data, utilizing UHF and SHF bands.

Scatter systems operate in the UHF and SHF bands to provide multi-channel telephony links.

At UHF and higher frequencies, radio signals can be made extremely directional; antenna systems using large parabolic reflectors produce very narrow radio beams, just as searchlight reflectors produce powerful beams of light. Microwave "dishes" are often used for multi-channel services, both terrestrial and to satellites.

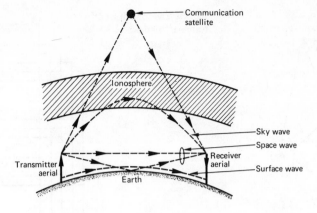

Fig. 3.1 Radio propagation methods

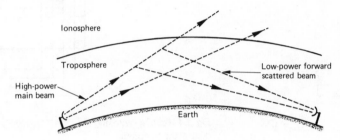

Fig. 3.2 Scatter propagation

4 Switching System Principles

A switching system of some sort is needed to enable any terminal (e.g. a telephone, a teleprinter, a facsimile unit) to pass information to any other terminal, as selected by the calling customer.

If the network is small, direct links can be provided between each possible pair of terminals and a simple selecting switch installed at each terminal (fig. 4.1). If there are 5 terminals, each must be able to access 4 links, so if there are N terminals there must be a total of $\frac{1}{2}N(N-1)$ links.

A slightly different approach would be to have one link permanently connected to each terminal, always used for calls to that particular terminal (fig. 4.2). Again, each terminal would need a selection switch to choose the distant end wanted for a particular conversation, but the number of links is reduced from 10 to 5 for a 5-terminal system, and to N links for N terminals.

As numbers of terminals and distances increase, this type of arrangement becomes impossibly expensive with today's technologies. A variant of this is however in wide use in radio-telephone networks: all terminals use a single common channel to give the instructions for setting up each

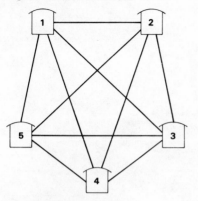

Fig. 4.1 Full interconnection

Fig. 4.2 The "one link per terminal" method

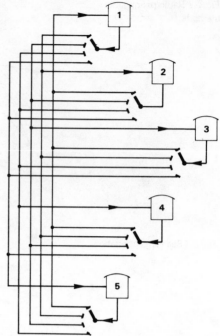

Fig. 4.3 Use of a calling channel

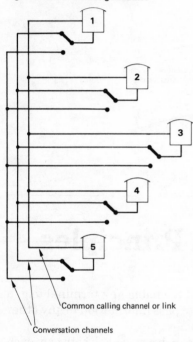

Common calling channel or link

Conversation channels

Fig. 4.4 Use of a switching centre: the telephone exchange

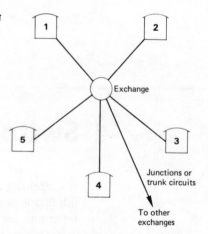

Exchange

Junctions or trunk circuits

To other exchanges

call (fig. 4.3). The terminals concerned then both switch to the allocated link or channel for their conversation. The number of links may be reduced substantially by this method; enough need be provided only to carry the traffic generated by the system. But each terminal still needs access to several channels and must have its own selection switch.

So far as telephone networks are concerned it is at present more economical to perform all switching functions at central points (i.e. not to make each terminal do its own switching). This means the provision of only one circuit from the nearest switching point or exchange to each subscriber's terminal (see fig. 4.4). As networks grow and expand into other areas, more switching points are provided, with special circuits between such points to carry the traffic between the areas concerned. The whole network has to be designed to strict transmission parameters so as to ensure that any subscriber on any exchange can converse satisfactorily with all other subscribers, anywhere in the world. Call accounting equipment also has to be provided so that appropriate charges can be levied for all calls made.

The first telephone exchanges were manual and all calls were established by operators. Automatic exchanges using electromechanical relays and switches were however developed very rapidly. Computer-controlled exchanges, with no moving parts at all, are now beginning to become common. Most countries in the world now have automatic telephone systems fully interconnected with the rest of the world's systems.

Part B Some Fundamentals

5 The Decibel

In telecommunication engineering, an engineer is concerned with the transmission of intelligence from one point to another, the intelligence being transmitted in the form of electrical signals. A telecommunication system which carries such signals may consist of a number of links in tandem and, certainly, each link will consist of a number of different items, such as transmission lines and amplifiers, also connected in tandem. Each item will introduce a certain loss, or gain, of power into the system.

Consider as a simple example a link of four items:

Item 1 introduces a loss with ratio $\dfrac{\text{Power out}}{\text{Power in}} = \dfrac{1}{2}$

Item 2 introduces a further loss with ratio $\dfrac{\text{Power out}}{\text{Power in}} = \dfrac{1}{50}$

Item 3 introduces a further loss with ratio $\dfrac{\text{Power out}}{\text{Power in}} = \dfrac{1}{10}$

Item 4 introduces a gain with ratio $\dfrac{\text{Power out}}{\text{Power in}} = \dfrac{10\,000}{1}$

Then, if there are no complications in the circuitry, the overall ratio is:

$$\frac{\text{Power out}}{\text{Power in}} = \frac{1}{2} \times \frac{1}{50} \times \frac{1}{10} \times \frac{10\,000}{1} = 10$$

In other words the signal will be ten times as strong at the end as it was at the beginning.

In practice we do not often have such friendly numbers to deal with. The multiplication of a whole chain of numbers like

$$0.001347 \times 0.0418 \times 0.1117 \times 2174.32$$

would, however, be somewhat tedious, and the power ratios we deal with in telecommunications work invariably involve inconveniently large or very small numbers.

To multiply numbers together you do of course have to add together the logarithms of these numbers. Since adding is much easier to do than multiplying, it was decided to concentrate on the logarithms of power ratios rather than the ratios themselves.

This unit, the logarithm of the ratio of in and out powers, is called a Bel, after the Scottish-born American inventor of the telephone, Graham Bell. The Bel turned out to be too large a unit for normal everyday use so it has been divided into ten to become the decibel or dB.

$$\text{Number of dB} = 10 \log_{10} \frac{\text{Power out}}{\text{Power in}}$$

If half the input power is lost, there is a loss of 3 dB. If an amplifier boosts power by 10 000 times, there is a gain of 40 dB. The decibel is widely used in all telecommunication disciplines, and it is not as complicated as it seems. If, for example, a type of cable introduces 1 dB of loss per kilometre, a 6.0 kilometre length will introduce a total loss of 6.0 dB. If this is followed by two circuits, one giving a loss of 3 dB and the other a gain of 15 dB, the total effect is obtained by plain addition, an overall gain of 6 dB.

The human ear is capable of responding to a wide range of sound intensities and has a sensitivity which varies with change in amplitude in a logarithmic manner. This makes the decibel a convenient unit for use with sound measurement and the measurement of sound equipment also; it is for example used when measuring the noise of aircraft taking off.

The decibel is not an absolute unit but is only a measure of a power ratio. It is meaningless to say, for example, that an amplifier has an output of 60 dB unless a reference level is quoted or is clearly understood. For example, a 60 dB increase on 1 microwatt gives a power level of 1 watt and a 60 dB increase on 1 watt gives a power level of 1 megawatt. Here the same 60 dB difference expresses power differences of less than 1 watt in one case and nearly one million watts in the other. It is therefore customary in telecommunication engineering to express power levels as so many decibels above, or below, a clearly understood reference power level. This practice makes the decibel a more significant unit and allows it to be used for absolute measurements. The reference level most commonly used is 1 milliwatt, and a larger power, P_1 watts, is said to have a level of

$$+x \text{ dBm} \quad \text{where } x = 10 \log_{10} \left(\frac{P_1}{1 \times 10^{-3}} \right)$$

and a smaller power, P_2 watts, is said to have a level of

$$-y \text{ dBm} \quad \text{where } y = 10 \log_{10} \left(\frac{1 \times 10^{-3}}{P_2} \right)$$

6 Frequency, Waveforms and Filters

1 *Direct and Alternating Current*

When an electric current is sent along a wire, energy is dissipated as heat. The amount of heat generated in the wire is energy sent out from the source but lost on the way and not available for use at the distant end; such energy is dependent on the square of the current which is flowing. This means for example that, if the current goes up by a factor of 5, the energy lost on the way goes up by a factor of 25. In some circumstances losses on this scale might well be unacceptable.

Electric power stored in batteries produces a steady voltage which will drive a steady current in one direction, called direct current or d.c., round a circuit and back to the other terminal of the battery.

To transfer large quantities of energy from power stations to distribution points, it is clearly desirable for as low a current as possible to flow, to minimize heat losses. One basic fact about electricity is that the same total amount of power can be sent if the voltage is multiplied by a factor, say 100, and the current divided by the same factor. Heat dissipation on the way (proportional to the square of the current) would then drop to 1/10 000th part of what it was before—or it might be economic to use a thinner and cheaper gauge of wire and accept a reduced heat-loss saving.

Current which can be made to reverse direction at regular intervals is called alternating current, a.c. Alternating current can be fed to a transformer which enables its voltage levels to be changed either up or down to suit particular requirements. The use of a.c. at high voltage enables large amounts of power to be transmitted across a country, without huge heat losses; these high voltages are then transformed down to the voltage levels we use domestically in our houses.

(High d.c. voltages are just as efficient for transferring power with reduced heat losses and are used in some specialized applications. The generation of a.c. is however somewhat simpler than that of d.c., and only a.c. can be fed through transformers to change the supply voltage down from the extremely high voltages used for transmission to levels safe enough to be used in homes.)

One simple form of alternating voltage generator (called an alternator) is shown in fig. 6.1. This is a loop of wire rotated in a magnetic field. The voltage induced in a conductor moving in a magnetic field is proportional to the rate at which the moving wire cuts the magnetic flux. In fig. 6.1a the conductors AB and CD cut the magnetic field between the North and South magnetic pole pieces at right angles, so the maximum or peak voltage is induced in the coil. When the wire loop has rotated through one quarter turn to that in fig. 6.1b, both conductors are moving parallel to the magnetic field and so are not cutting the flux at all and there is zero induced voltage. Fig. 6.2 shows how the induced electromotive force or

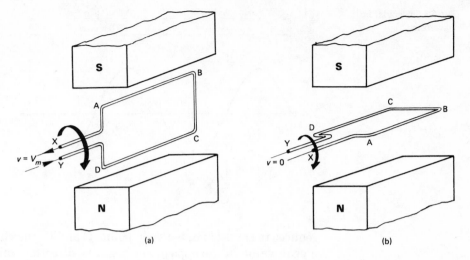

Fig. 6.1 A single loop alternator: (*a*) conductors cutting the flux at the maximum rate to produce the maximum induced voltage, (*b*) in this case the conductors do not cut the flux and the induced e.m.f. is zero

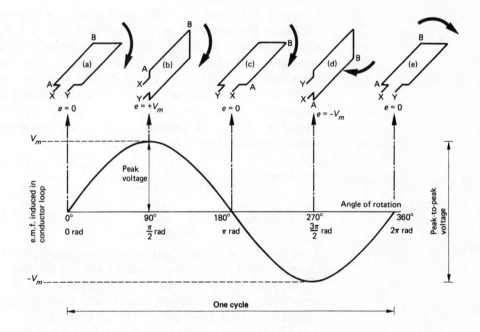

Fig. 6.2. One complete cycle of a sinusoidal wave form

voltage varies over a complete revolution of the loop, providing one cycle of the alternating voltage.

2   *The Sinusoidal Waveform*

Many waveforms are possible with alternating currents. One of the simplest to produce comes by the rotation of a loop of wire in a uniform magnetic field as described. This is called a sinusoidal waveform and is shown in fig. 6.3.

Between points A and B the current increases from zero to a peak value in the positive direction. Between points B and C the current gradually

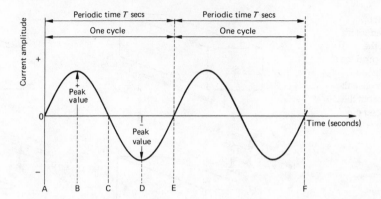

Fig. 6.3 Sinusoidal a.c. waveform

reduces to zero. Then, between points C and D, the current "increases" to a peak value in the opposite or negative direction, and between points D and E it gradually "reduces" to zero again. This whole sequence from point A to point E represents one complete rotation of the wire loop in the magnetic field, and is called a cycle of a.c. waveform.

Clearly the cycle is repeated between points E and F, representing another rotation of the wire loop, and this waveform is repeated for each subsequent rotation. The time needed, in seconds, for one cycle of waveform to be produced is called the periodic time T of the a.c. waveform.

The number of complete cycles occurring in one second is called the frequency f of the a.c. waveform in hertz (Hz). One hertz is one cycle per second. From fig. 6.4 it should be clear that frequency and periodic time are reciprocals of each other. That is

$$\text{Frequency} = \frac{1}{\text{Periodic time}}$$

$$\text{Periodic time} = \frac{1}{\text{Frequency}}$$

with frequency in hertz and time in seconds. In fig. 6.4, for example, the frequency is 4 Hz and the periodic time is $\frac{1}{4}$ second.

The strength of the current at any instant in time is called the amplitude of the waveform, and the direction of the current (positive or negative) is called the polarity of the current.

It will also be clear from fig. 6.3 and fig. 6.4 that the amplitude reaches a peak value in the positive and negative directions once every cycle.

We have, as one example, already associated the production of a sinusoidal waveform with the rotation of a loop of wire in a magnetic field, and the resulting current plotted against time, as in fig. 6.3 and fig. 6.4. We could also consider the loop as moving through 360° in one rotation, so we could plot the resultant current against angular rotation, as shown in fig. 6.5.

Also, if we consider the energy of an a.c. waveform travelling through space or along a transmission line at a particular velocity, then a certain distance will be travelled in the periodic time for one cycle, as shown in fig. 6.6. We can see now that the a.c. waveform repeats complete cycles

Fig. 6.4 Sinusoidal a.c. waveform with a frequency of 4 Hz

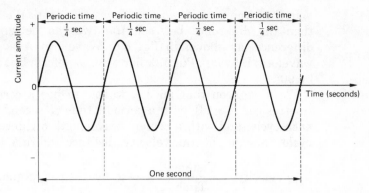

Fig. 6.5 Sinusoidal a.c. waveform plotted against angular rotation

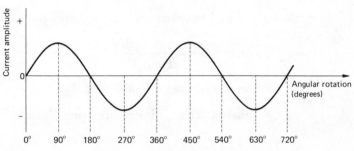

Fig. 6.6 Sinusoidal a.c. waveform plotted against distance

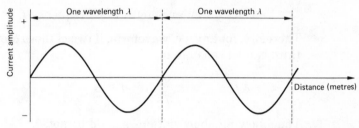

Fig. 6.7 Sinusoidal a.c. waveform related to phase

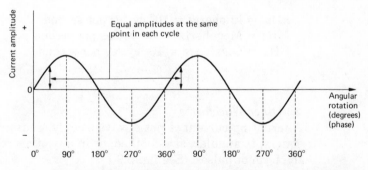

over equal distances. The distance representing each cycle is called the wavelength of the a.c. waveform in metres. The Greek letter lambda λ is used as the symbol for wavelength.

In fig. 6.5 the rotation of the loop is shown as a continuously increasing number of degrees. Alternatively, we could consider the start of each rotation as beginning from 0°, so each successive cycle in fact occurs from 0 to 360°, as shown in fig. 6.7. Clearly, at the same point in each cycle, the amplitude of the waveform has the same value. This way of identifying a particular point in any cycle as a degree of rotation is called the phase of the a.c. waveform.

In the same way, if two waveforms are identical except for their phase, then the difference between the two can be expressed as a phase difference, as shown in fig. 6.8. Waveform A is seen to be leading waveform B by 90°. Put another way, waveform B is lagging waveform A by 90°.

We have seen that an a.c. waveform has a certain energy velocity (metres per second), with a periodic time (T seconds) for the duration of each cycle, and with a certain wavelength distance (λ metres) for each cycle. Now, in general, velocity, distance and time are related by

$$\text{Velocity} = \frac{\text{Distance}}{\text{Time}} \quad \text{(e.g. metres per second, km/h)}$$

So for any a.c. waveform of wavelength λ and periodic time T,

$$\text{Velocity } v = \frac{\text{Wavelength } \lambda}{\text{Time } T}$$

But it has already been established that

$$\text{Frequency } f \text{(Hz)} = 1/\text{Periodic time } T \text{ (secs)}$$

so that

$$\text{Velocity } v = \text{Wavelength } \lambda \times \text{Frequency } f$$

Therefore, for any a.c. waveform, if two of these properties are known, the third can be calculated:

$$v = \lambda f \qquad \lambda = \frac{v}{f} \qquad f = \frac{v}{\lambda}$$

The following abbreviations should be noted:

kHz = kilohertz = 10^3 cycles per second
MHz = Megahertz = 10^6 cycles per second
GHz = Gigahertz = 10^9 cycles per second

3 *Other Waveforms*

Alternating current is not always exactly sinusoidal: waveforms can be square, triangular, saw-toothed, or in any other repetitive shape, e.g. in the form of pulses. (See fig. 6.9.)

A mathematician called Fourier proved that any recurrent waveform of frequency F can be resolved into the sum of a number of sinusoidal waveforms having frequencies F, $2F$, $3F$, etc. In other words, any steady signal may be split up into a fundamental (frequency F) and harmonics (multiples of the fundamental frequency F, i.e. $2F$, $3F$, etc). Sounds produced by human voices are nearly always rich in harmonics.

A completely square waveform may, for example, be reconstructed by adding together the fundamental and an infinite series of odd harmonics. In fig. 6.10, a close approximation to a square wave is produced by adding to the fundamental only the 3rd and 5th harmonics.

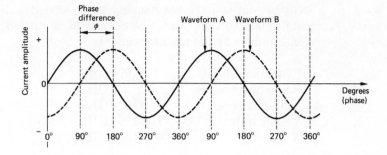

Fig. 6.8 Phase difference between two waveforms

Fig. 6.9 Alternating voltages with square and triangular waveforms

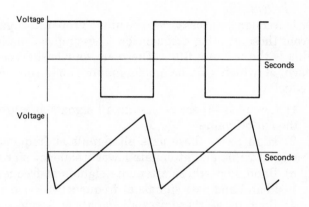

Fig. 6.10 Addition of fundamental sine wave, with third and fifth harmonics, to give an approximation to a square waveform

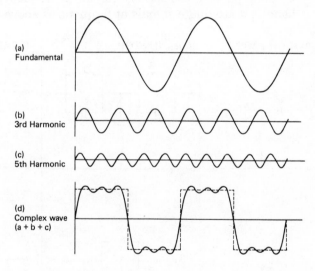

4 *Frequency Ranges*

Typical commonly encountered frequencies are

Human voice 100 Hz to 10 000 Hz
Human hearing 20 Hz to 15 000 Hz
Commercial speech 300 to 3400 Hz
Mains electricity 50 or 60 Hz.

If you hear mains-driven electric equipment hum quietly when you switch it on, you are not hearing the electricity; the humming noise is usually due to components being moved or rattled or changing dimensions slightly at the mains frequency.

Radio (broadcast)	1000 kHz
Radio (microwave)	6 GHz
Infra-red rays	100 000 GHz
Visible light	1 000 000 GHz
X-rays	100 million GHz

5 Filters

It is frequently necessary to be able to separate signals at one frequency from those at other frequencies. The simplest method of accomplishing this is the use of a filter. There are four basic types of filter (fig. 6.11), each of which can be made up in two forms using capacitors and inductors.

1) Low pass—these attenuate all signals at a greater frequency than the cut-off value.

2) High pass—these pass all signals at frequencies higher than the cut-off value and attenuate lower frequency signals.

3) Band stop—these attenuate signals at frequencies within a specified band and pass signals at frequencies above and below this band.

4) Band pass—these pass all signals at frequencies within a specified band and attenuate signals at frequencies above and below the band.

Fig. 6.11 Basic filters

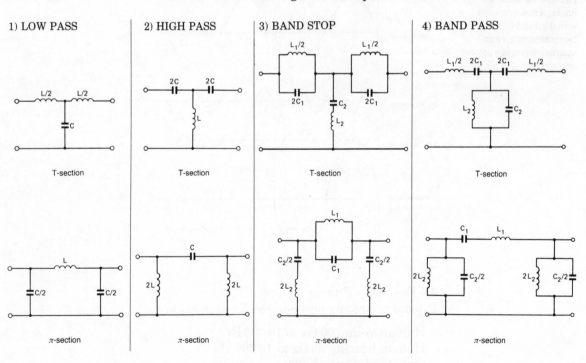

1) LOW PASS	2) HIGH PASS	3) BAND STOP	4) BAND PASS

7 Voice Frequencies

When we speak, our vocal cords vibrate and the resulting sounds, which are rich in harmonics but of almost constant pitch, are carried to the cavities in the mouth, throat and nose. Here the sounds are given some of the characteristics of the desired speech by the emphasizing of some of the harmonics contained in the sound waveforms and the suppression of others. Sounds produced in this way are the vowels, a, e, i, o and u, and contain a relatively large amount of sound energy. Consonants are made with the lips, tongue and teeth and contain much smaller amounts of energy and often include some relatively high frequencies.

The sounds produced in speech contain frequencies which lie within the frequency band 100–10 000 Hz. The pitch of the voice is determined by the fundamental frequency of the vocal cords and is about 200–1000 Hz for women and about 100–500 Hz for men.

The power content of speech is small, a good average being of the order of 10–20 microwatt. However, this power is not evenly distributed over the speech-frequency range, most of the power being contained at frequencies in the region of 500 Hz for men and 800 Hz for women.

The notes produced by musical instruments occupy a much larger frequency band than that occupied by speech. Some instruments, such as the organ and the drum, have a fundamental frequency of 50 Hz or less, while many other instruments, for example the violin and the clarinet, can produce notes having a harmonic content in excess of 15 000 Hz. The power content of music can be quite large. A sizeable orchestra may generate a peak power somewhere in the region of 90–100 watts while a bass drum well thumped may produce a peak power of about 24 watts.

When sound waves are incident upon the ear they cause the ear drum to vibrate. Coupled to the ear drum are three small bones which transfer the vibration to a fluid contained within a part of the inner ear known as the cochlea. Inside the cochlea are a number of hair cells, and the nerve fibres of these are activated by vibration of the fluid. Activation of these nerve fibres causes them to send signals, in the form of minute electric currents, to the brain where they are interpreted as sound.

The ear can only hear sounds whose intensity lies within certain limits; if a sound is too quiet it is not heard and, conversely, if a sound is too loud it is felt rather than heard and causes discomfort or even pain. The minimum sound intensity that can be detected by the ear is known as the "threshold of hearing or audibility" and the sound intensity that just produces a feeling of discomfort is known as the "threshold of feeling". The ear is not, however, equally sensitive at all frequencies, as shown in fig. 7.1. In this diagram, curves have been plotted showing how the thresholds of audibility and feeling vary with frequency for an average person.

It can be seen that the frequency range over which the average human ear is capable of responding is approximately 30–16 500 Hz, but this

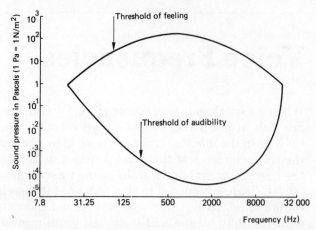

Fig. 7.1 The thresholds of audibility and feeling

range varies considerably with the individual. The ear is most sensitive in the region of 1000 to 2000 Hz and becomes rapidly less sensitive as the upper and lower limits of audibility are approached. The limits of audibility are clearly determined not only by the frequency of the sound but also by its intensity. At the upper and lower limits of audibility, the thresholds of audibility and feeling coincide and it becomes difficult for an observer to distinguish between hearing and feeling a sound.

In an ideal telecommunications system, all the frequencies present in a speech or music waveform would be converted into electrical signals, transmitted over the communication system, and then reproduced as sound at the distant end. In practice, this is rarely the case, for two reasons. Firstly, for economic reasons the devices used in circuits that carry speech and music signals have a limited bandwidth; secondly, particularly for the longer-distance routes, a number of circuits are often transmitted over a single telecommunication system and this practice provides a further limitation of bandwidth. It is thus desirable to have some idea of the effect on the ear when it is responding to a sound waveform, the frequency components of which have amplitude relationships differing from those existing in the original sound.

By international agreement the audio-frequency band for a "commercial quality" speech circuit routed over a multi-channel system is restricted to 300–3400 Hz. This means that both the lower and upper frequencies contained in the average speech waveform are not transmitted. To the ear, the pitch of a complex, repetitive sound waveform is the pitch corresponding to the frequency difference between the harmonics contained in the waveform, i.e. the pitch is that of the fundamental frequency. Hence, even though the fundamental frequency itself may have been suppressed, the pitch of the sound heard by the listener is the same as the pitch of the original sound. However, much of the power contained in the original sound is lost. Suppression of all frequencies above 3400 Hz reduces the quality of the sound but does not affect its intelligibility. Since the function of a telephone system is to transmit intelligible speech, the loss of quality can be tolerated; sufficient quality remains to allow a speaker's voice to be recognized.

Music is, however, badly distorted when transmitted over a normal telephone line because both low and high frequency notes are lost.

If a telecommunications link is required to carry a television signal, an even wider bandwidth needs to be transmitted without distortion.

A rough rule of thumb for the bandwidths needed for acceptable transmission is:

Voice   4 kHz
Music   10 kHz–15 kHz
Colour TV   8 MHz

8 Attenuation and Noise

As an information signal travels along a line its amplitude or power is progressively reduced because of losses in the line. These losses, called attenuation, are of two types:

1) Losses due to heat dissipation caused by the resistance of the conductors and the insulation resistance between the conductors.
2) Dielectric losses which affect alternating current (or a.c.) only and are dependent on the dimensions and type of insulant between the conductors themselves and between conductors and earth.

Attenuation generally increases as the frequency of the information signal increases; this variation is called attenuation distortion.

Fig. 8.1 shows how the attenuation of ordinary audio-frequency cable pairs of different gauges (diameters of 0.9 and 0.63 mm) varies with frequency. As you would expect, the thicker the wire the lower the attenuation. A telephone cable serving a subscriber many kilometres from the exchange may therefore have to be made up of heavier gauge conductors

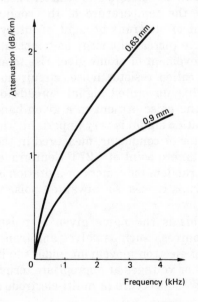

Fig. 8.1 Attenuation/ frequency characteristics of two typical audio-frequency cables

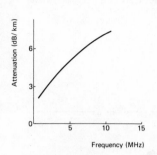

Attenuation (dB/ km)

Frequency (MHz)

Fig. 8.2 Attenuation/
frequency characteristics
of a coaxial pair

than those needed to serve phones in an office just next door to the exchange. It is not uncommon for there to be a loss of 10 dB between a subscriber and the subscriber's own exchange; this means that only 1/10th of the transmitted power of the original voice signal reaches the local exchange. Similar losses could be experienced at the other end also, so even an own-exchange call can often mean that only 1/100th of the original power is available: a 20 dB loss. Even with 30 dB of attenuation (1/1000th of the power), speech is usually still possible, so long as there isn't too much noise on the line.

The attenuation suffered by a signal passing along an audio-frequency cable like this continues to increase at frequencies above those plotted in the figure and could be as high as 30 dB/km at a typical carrier frequency at 1 MHz. For comparison, the attenuation/frequency characteristics for a coaxial cable, designed to be used for carrier telephony at high frequencies, are given in fig. 8.2.

In any telecommunication system, whether using line or radio links, there is unwanted electrical energy present as well as that of the wanted information signal. This unwanted electrical energy is generally called noise and arises from a number of different sources, which will now be considered very briefly.

1) *Resistor noise* A conductor is designed to carry current with minimum opposition, consistent with size and cost.

A resistor is a component designed to have a particular opposition to the flow of electrical current in a particular circuit. This opposition is called resistance in d.c. circuits, but in a.c. circuits the term impedance is used because of other, frequency-dependent, factors. In either case the unit used is the ohm (symbol Ω).

An electric current is produced by the movement of electrons dislodged by an externally applied voltage from the outer shells of the atoms making up the conductor material or resistor material. The movement or agitation of atoms in conductors and resistors is somewhat random, and is determined by the temperature of the conductor or resistor. The random movement of electrons brought about by thermal agitation of atoms tends to have increased energy as temperature increases.

This random movement of atoms gives rise to an unwanted electrical voltage which is called resistor noise, circuit noise, Johnson noise or thermal noise. This unwanted signal spreads over a wide range of frequencies, and the noise present in a given bandwidth required for a particular information signal is very important. This is the noise temperature of the resistor or conductor, measured in the Kelvin temperature scale, which has its zero point at $-273°$ Centigrade. This is the temperature at which the random movement or agitation of atoms in conducting or resistive materials ceases, so unwanted noise voltages are therefore zero.

2) *Shot noise* This is the name given to noise generated in active devices (energy sources), such as valves and transistors, by the random varying velocity of electron movement under the influence of externally applied potentials or voltages at appropriate terminals or electrodes.

3) *Partition noise* This occurs in multi-electrode active devices such as

transistors and valves and is due to the total current being divided between the various electrodes.

4) *Fluctuation noise* This can be natural (electric thunderstorms, etc.) or manmade (car ignition systems, electrical apparatus, etc.) and again spreads over a wide range of frequencies. Such noise can be picked up by active devices and conductors forming transmission lines.

5) *Static* This is the name given to noise encountered in the free-space transmission paths of radio links, and is due mainly to ionospheric storms causing fluctuations of the earth's magnetic field. This form of noise is affected by the rotation of the sun and by the sunspot activity that prevails.

6) *Cosmic or Galactic noise* This type of noise is also most troublesome to radio links, and is mainly due to nuclear disturbances in all the galaxies of the universe.

7) *Crosstalk* In multi-pair cables there is capacitive and inductive coupling between different pairs which produces an unwanted noise signal on any pair because signals are transmitted from other pairs. This is called crosstalk and can be reduced to some extent by twisting the conductors of each pair, or by changing the relative positions of pairs along the cable during manufacture, or by balancing the pairs over a particular route after installation.

8) *1/f noise* Fluctuations in the conductivity of the semiconductor material produce a noise source which is inversely proportional to frequency. This type of noise is also known as current noise, excess noise, or flicker noise.

In any telecommunications sytem, therefore, there will be a certain level of noise power arising from all or some of the sources described, with the noise power generally being of a reasonably steady mean level, except for some noise arising from impulsive sources such as car ignition systems and lightning. Noise which has a sensibly constant mean level over a particular frequency bandwidth is generally called white noise.

The presence or absence of unwanted noise on a circuit is one of the ways in which the quality of a circuit can be described. The received level is measured when a wanted signal of specified power level is present. This is compared with the received level measured when there is no signal present. This enables the ratio

$$\frac{\text{Signal plus Noise}}{\text{Noise}}$$

to be calculated. It is usually expressed in decibels and called the signal-to-noise ratio. The higher the ratio, the higher the quality of the circuit.

9 Amplitude Modulation

We have seen from section 7 that a bandwidth of from 300 Hz to 3400 Hz is required for the transmission of commercial quality speech.

To economize on cable it is desirable to be able to transmit more than one conversation over a single pair of wires. If several conversation signals were all connected together at one end of a line, it would not be possible to separate them at the distant end since each conversation would be occupying the same frequency spectrum of 300 Hz to 3400 Hz. Amplitude modulation (AM) plus frequency division multiplexing (FDM) is one way of solving this problem. Each conversation is shifted to a different part of the frequency spectrum by using a high-frequency waveform to "carry" each individual speech signal. These high frequencies are called carrier frequencies.

Amplitude modulation is the process of varying the amplitude of the sinusoidal carrier wave by the amplitude of the modulating signal, and is illustrated in fig. 9.1.

The unmodulated carrier wave has a constant peak value and a higher frequency than the modulating signal but, when the modulating signal is applied, the peak value of the carrier varies in accordance with the instantaneous value of the modulating signal, and the outline waveshape or "envelope" of the modulated wave's peak values is the same as the original modulating signal waveshape. The modulating signal waveform has been superimposed on the carrier wave.

It can be shown by mathematical analysis that, when a sinusoidal carrier wave of frequency f_c Hz is amplitude-modulated by a sinusoidal modulating signal of frequency f_m Hz, then the modulated carrier wave contains three frequencies.

One is the original carrier frequency, f_c Hz.

The second is the sum of carrier and modulating signal frequencies, $(f_c + f_m)$ Hz.

The third is the difference between carrier and modulating signal frequencies, $(f_c - f_m)$ Hz.

This is illustrated in fig. 9.2.

It should be noted that two of these frequencies are new, being produced by the amplitude-modulation process, and are called side-frequencies.

The sum of carrier and modulating signal frequencies is called the upper side-frequency. The difference between carrier and modulating signal frequencies is called the lower side-frequency. This is illustrated in the frequency spectrum diagram of fig. 9.3.

The bandwidth of the modulated carrier wave is

$$(f_c + f_m) - (f_c - f_m) = 2f_m$$

i.e. double the modulating signal frequency.

When the modulating signal consists of a band of frequencies, as already seen for commercial speech and music for example, then each

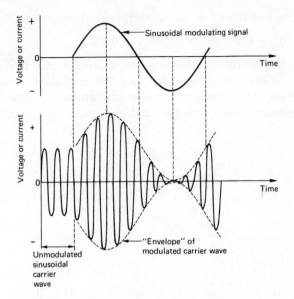

Fig. 9.1 Amplitude-modulated carrier wave

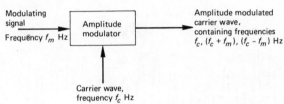

Fig. 9.2 Principle of amplitude modulation

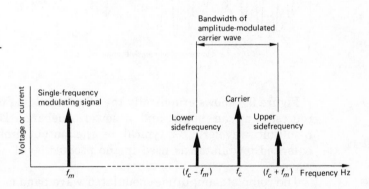

Fig. 9.3 Frequency spectrum of an amplitude-modulated wave for single-frequency modulating signal

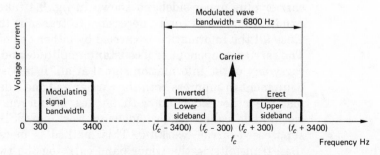

Fig. 9.4 Frequency spectrum of an amplitude-modulated wave for commercial speech modulating signal

individual frequency will produce upper and lower side-frequencies about the unmodulated carrier frequency, and so upper and lower sidebands are obtained. This is illustrated in fig. 9.4.

The bandwidth of the modulated carrier wave is

$$(f_c + 3400) - (f_c - 3400) = 6800 \text{ Hz}$$

which is double the highest modulating signal frequency.

It follows therefore that, as the modulating signal bandwidth increases, the modulated wave bandwidth also increases, and the transmission system used must be capable of handling this bandwidth throughout.

Fig. 9.5 Addition of two sidebands (without carrier)

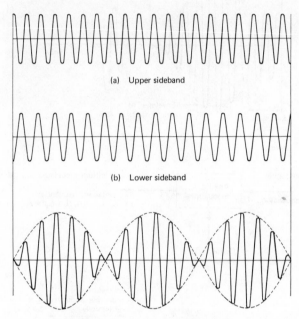

(a)   Upper sideband

(b)   Lower sideband

(c)   Upper and lower sideband present simultaneously

Figure 9.5 shows graphically the effect of adding together two signals, representing an upper and a lower sideband. The envelope of the resultant waveform is typical of the output from various types of balanced modulator as used in the production of amplitude-modulated waves.

The complete amplitude-modulated wave band of lower sideband plus carrier plus upper sideband shown in fig. 9.4 takes up more frequency bandwidth than is really necessary to transmit the information signal since all the information is carried by either one of the sidebands alone. The carrier component is of constant amplitude and frequency so does not carry any of the information signal at all. It is possible by using special equipment to suppress both the carrier and one sideband and to transmit just the other sideband with no loss of information. This method of working is called single sideband working (SSB), or single-sideband suppressed-carrier working. This method is more costly and complex than transmitting the wider band carrying the two sidebands plus the carrier and is not used for domestic radio broadcasting, but it is used for some long-distance radio telephony systems and for multi-channel carrier systems used in national telephone networks.

10 Frequency Modulation

Another method of superimposing information signals on to a carrier signal is frequency modulation in which the modulating signal varies the frequency of a carrier wave. This has a number of advantages over amplitude modulation. Frequency modulation is used for sound broadcasting in the VHF band, for the sound signal of 625-line television broadcasting, for some mobile systems, and for multi-channel telephony systems operating in the UHF band. The price which must be paid for some of the advantages of frequency modulation over double sideband amplitude modulation is a wider bandwidth requirement.

When a sinusoidal carrier wave is frequency modulated, its instantaneous frequency is caused to vary in accordance with the characteristics of the modulating signal. The modulated carrier frequency must vary either side of its nominal unmodulated frequency a number of times per second equal to the modulating frequency. The magnitude of the variation—known as the frequency deviation—is proportional to the amplitude of the modulating signal voltage.

The concept of frequency modulation can perhaps best be understood by considering a modulating signal of rectangular waveform, such as the waveform shown in fig. 10.1. Suppose the unmodulated carrier frequency is 3 MHz. The periodic time of the carrier voltage is $\frac{1}{3}\mu$s and so three complete cycles of the unmodulated carrier wave will occur in $1\,\mu$s. When, after $1\,\mu$s, the voltage of the modulating signal increases to $+1$ V, the instantaneous carrier frequency increases to 4 MHz. Hence in the time interval $1\,\mu$s to $2\,\mu$s, there are four complete cycles of the carrier voltage. After $2\,\mu$s the modulating signal voltage returns to 0 V and the instantaneous carrier frequency falls to its original 3 MHz. During the time interval $3\,\mu$s to $4\,\mu$s the modulating signal voltage is -1 V and the carrier frequency is reduced to 2 MHz; this means that two cycles of the carrier voltage occur in this period of time.

Fig. 10.1 A frequency-modulated wave

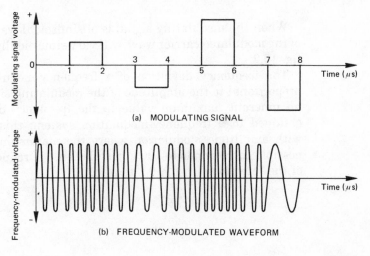

(a) MODULATING SIGNAL

(b) FREQUENCY-MODULATED WAVEFORM

When, after $4\,\mu s$, the modulating voltage is again $0\,V$, the instantaneous carrier frequency is restored to $3\,MHz$. At $t = 5\,\mu s$ the modulating voltage is $+2\,V$ and, since frequency deviation is proportional to signal amplitude, the carrier frequency is deviated by $2\,MHz$ to a new value of $5\,MHz$. Similarly, when the modulating voltage is $-2\,V$, the deviated carrier frequency is $1\,MHz$. At all times the amplitude of the frequency-modulated carrier wave is constant at $1\,V$, and this means that the modulating process does not increase the power content of the carrier wave.

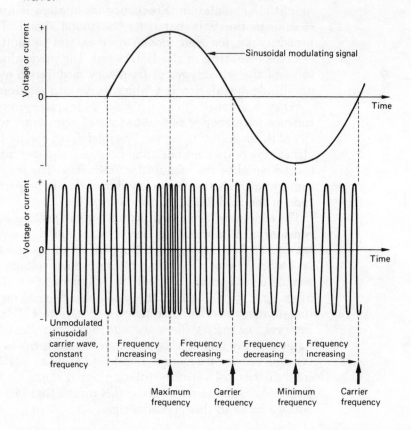

Fig. 10.2 A frequency-modulated carrier wave

When the modulating signal is of sinusoidal waveform, the frequency of the modulated carrier wave will vary sinusoidally; this is illustrated by fig. 10.2.

The frequency deviation of a frequency-modulated carrier wave is proportional to the amplitude of the modulating signal voltage. There is no inherent maximum value to the frequency deviation that can be obtained in a frequency-modulation system; this should be compared with amplitude modulation where the maximum amplitude deviation possible corresponds to 100% modulation, i.e. reduction of the amplitude of the "envelope" in fig. 9.1 to zero.

11 Pulse Modulation

Another method of conveying information is by means of pulses of voltage or current.

With pulse modulation the carrier wave is not sinusoidal, but consists of repeated rectangular pulses. The amplitude, width or position of the pulses can be altered by the information signal, as illustrated in fig. 11.1.

Fig. 11.1 **Pulse-modulated carrier waves**

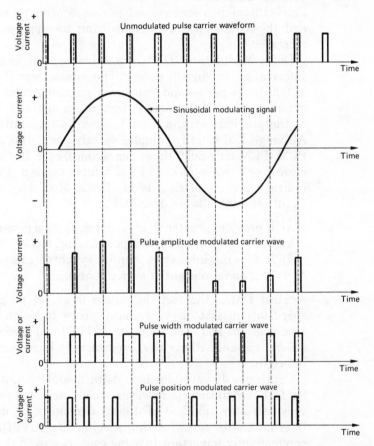

Pulse amplitude modulation is that form of modulation in which the amplitude of the pulse carrier is varied in accordance with some characteristic, normally the amplitude, of the modulating signal.

Pulse width modulation is that form of modulation in which the duration of a pulse is varied in accordance with some characteristic of the modulating signal. Sometimes pulse width modulation is called pulse duration modulation or pulse length modulation.

Pulse position modulation is that form of modulation in which the positions in time of the pulses are varied in accordance with some characteristic of the modulating signal without a modification of pulse width.

Pulse code modulation is a special form of pulse modulation and is dealt with in some detail in section 33; PCM can be considered to be the sampling of the modulating analog signal and the production of pulse amplitude modulated signals, i.e. pulse signals whose amplitudes are directly proportional to the amplitude of the original signal at the successive instants at which samples are measured. Each of these samples is then "quantized" or measured against a built-in scale of amplitudes (from 0 to 256 in modern systems); the number of this scale which represents the amplitude of each particular sample is then transmitted as a binary coded signal, i.e. a coded group of 8 bits (see section 32), to indicate the amplitude level ($2^8 = 256$). The signal in the line is therefore a series of pulses representing numbers. At the receiving end these numbers are decoded and an appropriate analogue signal recreated.

Delta Modulation (DM) is another special form of transmission of information by digital pulses. The analog signal is scanned, typically 32 000 times per second (but lower scanning rates can be used in some circumstances). If the signal's instantaneous value is greater than it was at the previous scan, a digit 1 signal is transmitted; if less, a digit 0. Although DM usually samples speech waveforms more frequently than PCM (PCM only 8000 times per second), each DM signal is only a single digit, 0 or 1, whereas each PCM sample needs a "PCM word" of 8 bits to indicate the quantizing level. Delta Mod does therefore have some significant advantages over PCM:

a) It provides greater channel capacity for a given bit-rate, resulting in higher pair-gain and lower per-channel costs.
b) It does not inherently require synchronization, as PCM does.
c) It is more tolerant of system noise.

Against DM is the fact that there is as yet no generally agreed DM specification. It took very many years to reach agreement on PCM specifications and even then the CCITT had to recognize two completely different quantization rules, one basically European, the other American in origin.

Continuously-variable-slope delta modulation, one of the contenders for DM recognition, is however widely accepted as being technically a very attractive method of digital transmission, despite its lack of an internationally agreed coding specification. CVSDM could well become commercially important in some countries even though it is a comparatively late arrival in the market place.

Part C Switching and Signalling

12 Step-by-Step Telephone Exchanges

Telephony was invented in the 1870s; all the early exchanges used human operators to establish and supervise calls. As networks grew it became uneconomic to continue to use people to set up telephone calls. It has indeed been calculated that to carry today's telephone traffic using 19th century practices would need more than half the total population of all major cities to be employed as telephone operators!

For automatic operation the first requirement is a way of indicating to the exchange the telephone number of the customer to whom you wish to speak. The rotary dial with ten finger holes is now nearly a century old in basic concept but is still in wide use. Contacts within the dial make and break an electrical circuit which interrupts current flowing, from a battery in the exchange, through the loop made by the line to the customer's premises and through the phone itself. If for example you dial 7, the dial breaks the circuit 7 times, with each break lasting a pre-determined time, usually about 1/20th of a second. Relays in the exchange respond to these break signals.

The step-by-step principle was the first automatic system to become practicable for public telephone exchanges; the selection of a particular line is based on a one-from-ten selection process. For example, fig. 12.1 shows a simple switch that has ten contacts arranged around a semicircular arc or bank, with a rotating contact arm or wiper that can be made to connect the inlet to any one of the ten bank contact outlets as required.

Fig. 12.1 Principle of switching by electro-mechanical uniselector of one-from-ten outlets

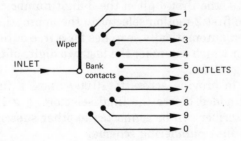

The wiper is rotated by a simple electro-magnet driving a suitable mechanism, so the arrangement is called an electro-mechanical switch.

The wiper rotates in one plane only, so this type of electro-mechanical switch is called a uniselector. Clearly the inlet can be connected to any one of the ten outlets, but the outlets are numbered from 1 to 0, which is normal practice in the step-by-step switching system.

This principle can be extended to enable the inlet to be connected to any one from 100 outlets by connecting each of the ten outlets of the first uniselector to the inlet of another uniselector, as shown in fig. 12.2. The switching of the inlet to any one of the 100 outlets (numbered 11 to 00) is done in two steps, the first digit being selected on the first uniselector, and the second digit being selected on the second uniselector.

If each of these 100 outlets is now connected to another uniselector, the inlet can then be connected to any one from 1000 outlets, numbered from 111 to 000, with the digits being selected one at a time on the three successive switching stages. This arrangement can theoretically be extended to accommodate any number of digits in a particular numbering scheme.

The same sort of numbering scheme can be provided (on a step-by-step basis) by a different type of electro-mechanical switch called a two-motion selector. The principle is illustrated simply in fig. 12.3a and b.

The bank of fixed contacts now contains 10 semi-circular arcs, each having 10 contacts, and arranged above each other. The moving contact or wiper can be connected to any one of the 100 bank contacts by first moving vertically to the appropriate level, and then rotating horizontally to a particular contact on that level. The 100 outlets are numbered from 11 to 00.

The diagram symbol used to illustrate the 100-outlet 2-motion selector is shown in fig. 12.3c.

As with the uniselector arrangement, the two-motion selector system can be extended to give access to any number of outlets by adding an extra switching stage for each extra digit required in the numbering scheme. A 3-digit numbering scheme from 111 to 000 outlets is illustrated in fig. 12.4 with the one-from-a-hundred selector preceded by a one-from-ten selector.

In fig. 12.4, the first digit of the 3-digit numbering scheme raises the wiper of the first 2-motion selector to the appropriate vertical level. The selector then automatically searches for a free outlet on that level to the next selector which caters for the last two digits of the 3-digit number, as shown in fig. 12.3a.

In order to provide access to 10 000 lines, a further stage of group selectors is added before the final selectors. Fig 12.5 illustrates how a calling subscriber can be connected to other subscribers in an exchange having a 4-digit numbering scheme.

Theoretically, a 4-digit numbering scheme can accommodate 10 000 subscribers, but it is necessary also to provide junctions to other exchanges, lines to the operator and other enquiry services, and so on. This means that the capacity of an exchange of this type would in practice be about 6000 subscribers instead of the theoretical 10 000.

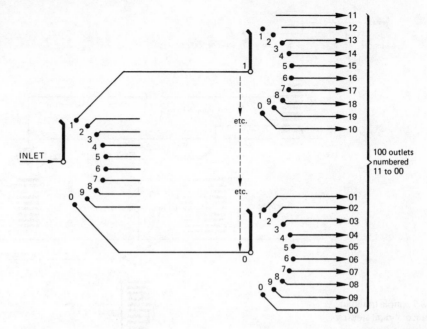

Fig. 12.2 Simple step-by-step selection of one-from-a-hundred outlets

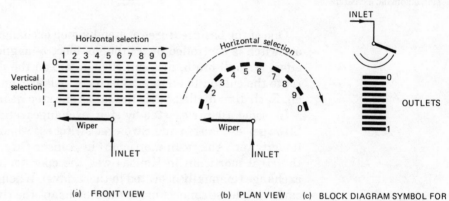

Fig. 12.3 Principle of one-from-a-hundred selection by two-motion selection

(a) FRONT VIEW   (b) PLAN VIEW   (c) BLOCK DIAGRAM SYMBOL FOR TWO-MOTION SELECTOR

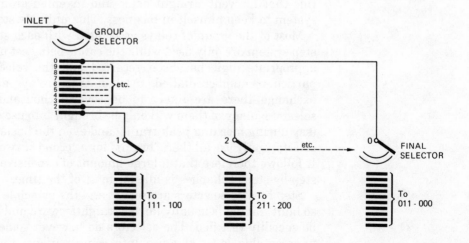

Fig. 12.4 Theoretical selection of one-from-a-thousand by two-stage step-by-step switching

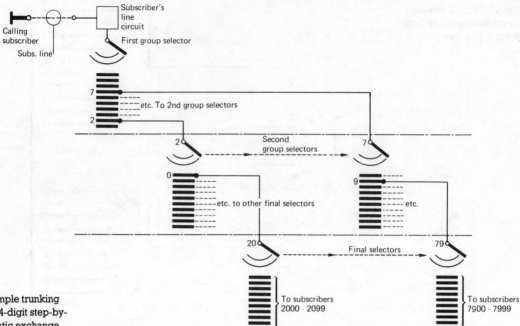

Fig. 12.5 Simple trunking diagram of 4-digit step-by-step automatic exchange

One of the basic features of step-by-step exchanges is that each selector or switch is controlled by a group of electro-magnetic relays which is in effect a small brain, just sufficient to act on the digit it receives and to route the call on to another selector which acts on the next digit, and so on. Each digit dialled takes the caller one step nearer the called number.

In some countries step-by-step exchanges and selectors are called Strowger exchanges and Strowger switches. Almon B. Strowger was an undertaker (American mortician) in Kansas City, Missouri. The wife of the rival mortician in Kansas was the operator in the local telephone exchange (manually operated in those days). Whenever an anguished call came in "please connect me to the mortician" the rival got the call and the funeral business. Almon B. Strowger was so incensed by the injustice of this that he went straight home and invented an automatic switching system to keep himself in business. This unlikely story really is true.

Most of the group of relays associated with each selector in a step-by-step system are only used while the call is being set up, so, as soon as the appropriate digit has been received and the selector stepped to the particular number dialled, most of these relays are idle. In a main exchange there are likely to be several thousand of these complex selectors; many of them will only be brought into use during busy periods (say during morning peak traffic) and even the busiest (or "first choice") selectors only use all their "brains" for a second or two every few minutes. It follows therefore that a large amount of expensive equipment in such step-by-step exchanges is idle for most of the time.

Step-by-step selectors are robust and the principles are easy to follow, so fault finding is usually fairly straightforward and faults can normally be speedily rectified. The selectors do however sometimes introduce an unacceptable level of noise into conversations; they "shudder" quite violently while calls are being set up and while they are releasing at the end of each call. These movements affect electrical resistance at contact points and so produce noise in circuits.

Selectors have a great many moving parts which means that regular lubrication, with cleaning of relay contacts and readjustment of switches from time to time, is absolutely essential if good service is to be maintained. The high cost of maintenance is one of the main reasons for the fact that electro-mechanical step-by-step exchanges are, in most countries, now being replaced by exchanges which use low-maintenance-cost electronic techniques.

13 Reed Relay and Crossbar Exchanges

A reed relay is a device based on the fact that an electric current passing through a coil of wire produces an electro-magnet, with the ends of the coil having opposite magnetic polarities, as in fig. 13.1.

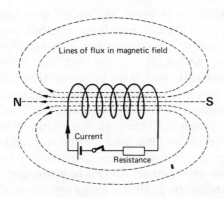

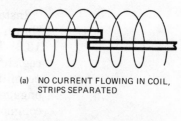

(a) NO CURRENT FLOWING IN COIL, STRIPS SEPARATED

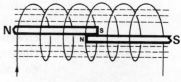

(b) CURRENT FLOWING IN COIL, STRIPS ARE MAGNETIZED AND ATTRACT EACH OTHER TO FORM AN ELECTRICAL CONTACT

Fig. 13.1 Coil of wire as a simple electro-magnet

Fig. 13.2 Principle of operation of a reed relay

If now two thin strips of material that can be magnetized are placed inside the coil, the strips will become magnetized when the current is flowing in the coil. If the two strips are placed so that one end of each overlaps, they will have opposite magnetic polarities and so will attract each other, as shown in fig. 13.2.

These two strips can be used to form a switch in another electrical circuit. The two strips are placed inside a glass envelope containing an inert gas, and the overlapping portions are coated with gold to give a good reliable electrical contact. The whole assembly contained by the glass envelope is called a reed insert, since it is placed inside the electro-magnet coil.

A typical reed relay has four of these reed inserts placed inside the electro-magnetic coil, each of which can be used to switch a separate electrical circuit. Coils are usually then arranged in a matrix formation so that the contacts in any particular reed relay in the matrix may be operated under the control of pulses of current through winding coils.

Selectors in crossbar exchanges have horizontal and vertical bars operated by electro-magnetic relay coils, so that, with a crossbar switch also, the contacts at a particular point in a matrix may be operated under the control of these relays.

Crossbar switches and reed relays are both used in telephone exchanges. The basic concept is however quite different from that of step-by-step exchanges.

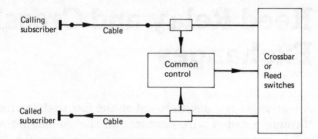

Fig. 13.3 Principle of reed relay and crossbar exchange

Instead of each switch or selector having its own little distributed "brain", there is a central "brain" which controls all switches (see fig. 13.3). This central "brain" or register/marker is rather like a computer; it registers the number dialled, it checks that the calling number is permitted to make the call, and tests to see if the called number is engaged. Exchanges using this centralised control function are called common control exchanges. If the called number is free, this common control equipment chooses a path through the exchange to join together the calling line to the called line, and issues instructions to all the crossbar switches or reed relays concerned to operate in such a way that the two lines are connected together. All this happens very rapidly—calls do not have to be switched through the exchange one digit at a time in step with the subscriber's dial.

This increased speed of operation means that it becomes attractive to replace ordinary rotary dials by push buttons; these allow calls to be set up very rapidly indeed.

As there is less mechanical complexity in a crossbar or reed relay system, it is more reliable than step-by-step. Also the path through the exchange is such that crossbar switches and reeds introduce less noise into a telephone conversation than step-by-step switches.

Crossbar switches and reed relays have no moving parts demanding regular maintenance and adjustment so very little routine maintenance is needed and fewer exchange men are required. It is this reduction of maintenance staff which usually results in large common control exchanges being more economic than similar size Strowger step-by-step exchanges. For small-capacity exchanges the extra cost of common-control equipment may be too great to be offset by maintenance savings

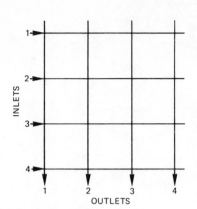

Fig. 13.4 Simple 4 × 4 switching matrix

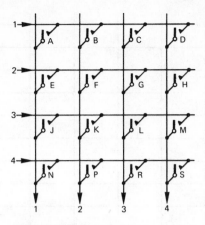

Fig. 13.5 Principles of switching by a 4 × 4 matrix switch

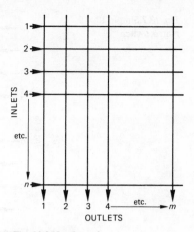

Fig. 13.6 Number of crosspoints in a matrix switch

so Strowger step-by-step type equipment is still widely used in some areas of the world.

Both crossbar and reed relay switching depend on the operation of a switching matrix, the principle of which can be explained by considering the circuits which are to be connected together as being arranged at right angles to each other in horizontal and vertical lines. These lines represent inlets and outlets of the switch. This idea is illustrated in fig. 13.4.

The intersections between horizontal and vertical lines are called crosspoints. At each crosspoint some form of switch contact is needed to complete the connection between horizontal and vertical lines, as shown in fig. 13.5. Any of the 4 inlets can be connected to any of the 4 outlets by closing the appropriate switch contacts. For example,

a) Inlet 1 can be connected to outlet 2 by closing contact B.
b) Inlet 4 can be connected to outlet 3 by closing contact R.

Considering figs 13.4 and 13.5 again, it can be seen that with 4 inlets and 4 outlets there are 16 crosspoints.

Obviously, the number of crosspoints in any matrix switch can be calculated by multiplying the number of inlets by the number of outlets. This is further illustrated in fig. 13.6. If there are n inlets and m outlets, then the number of crosspoints is $(n \times m)$.

a) If n is larger than m, that is if there are more inlets than outlets, then not all the inlets can be connected to a different outlet. When all the outlets have been taken, there will be some inlets still not in use.
b) If m is larger than n, that is there are more outlets than inlets, then, when all inlets are each connected to an outlet, there will be some outlets still not in use.

So, the maximum number of simultaneous connections that can be carried by a matrix switch is given by whichever of the number of inlets or outlets is smaller. For example, if there are 10 inlets and 5 outlets, then the maximum number of simultaneous connects possible is 5, as illustrated in fig. 13.7.

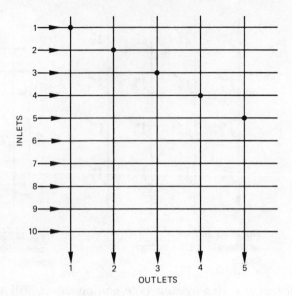

Fig. 13.7 Simple 10 × 5 matrix switch

Inlet 1 is connected to outlet 1
Inlet 2 is connected to outlet 2
Inlet 3 is connected to outlet 3
Inlet 4 is connected to outlet 4
Inlet 5 is connected to outlet 5
Inlet 6, 7, 8, 9, 10 cannot be used.

The efficiency in use of crosspoints may be calculated as

$$\frac{\text{Maximum number of crosspoints that can be used simultaneously}}{\text{Total number of crosspoints in one matrix}} \times 100\%$$

For a large matrix this efficiency is necessarily very low, e.g. a 15 × 15 matrix with 225 crosspoints is only able to use 15 of these at any one time, giving only 6.7% efficiency.

Efficiency can be improved by using smaller matrix switches, linked together; most crossbar and reed relay exchanges are designed on this basis, with a series of interconnected switches. This can, in some circumstances, lead to link congestion or internal blocking. Careful design of the exchange is needed in order to maximise its traffic handling capacity while minimizing equipment quantities (and therefore costs).

14 Stored Program Control

Control in telephone exchanges developed from individual control of each switch in step-by-step exchanges to the use of a small number of complex centralised units in so-called "common control" exchanges, which were mainly crossbar or reed. Common control units were originally completely electro-mechanical, using the same basic types of relay which had been used for many years in earlier exchanges. Component improvements, in particular the development of cheaper memories, rapidly led to the introduction of new designs of these control units.

The first electronic common control units made use of thermionic valves in place of "traditional" electro-mechanical relays. These units demonstrated principles but were never considered to be really serious contenders; electronic common control had to wait for the popularisation of the transistor and the printed circuit board before it became truly economic.

By the late 1960s computers were being developed very rapidly. It seemed to many of the brightest computer designers that they could readily use their computers as efficient and fully centralised common control units for telephone exchanges.

There were however a few snags. A general-purpose processor designed to perform efficiently in a business environment is not necessarily capable of performing satisfactorily all the duties required of a central processor operating in real time in a telephone exchange. The table below summarizes the main differences in design philosophies.

	General-purpose computer	Telephone exchange central processor
Complete stoppage	Inconvenient	Unacceptable
Faulty output	Unacceptable	Inconvenient

These inescapable factors led to the development by some telecommunications firms of wired logic units able to perform most of the required functions, while some firms developed their own computers, especially designed to perform telephone exchange control duties. Some manufacturers have concentrated on improving and developing efficient and powerful central processors alone, but in the last few years technological developments in the design of multi-function components seem to indicate a long-term trend in the telephone exchange field towards distributed control by microprocessors which refer to their parent main processors or to other micros for special functions or for facilities such as abbreviated dialling, which require access to large memories. Good signalling systems between these distributed processors could well, from a service point of view, prove just as important as the actual switching practice followed.

The use of computers to control telephone exchange switching is called Stored Program Control or SPC. This has been defined as the control of an automatic switching arrangement in which call processing is determined by a program stored in an alterable memory.

Stored program control techniques have been used for crossbar exchanges and reed relay exchanges. During the 1980s suitable solid state devices have however become available at competitive prices. Many major manufacturers now produce exchanges using completely solid state switching devices; there are no moving parts, nothing ever requires readjustment.

Until very recently the signals passing through all types of telephone exchange were analogue in nature, i.e. they were continuously variable in amplitude or frequency in response to changes of sound pressure impressed on the microphone at the speaking end of a circuit. All first-generation SPC exchanges were necessarily therefore exchanges which established physical, metallic, space-division switched paths for analogue signals although controlled by digital computers.

Now that the world is moving away from analog transmission and space-division switching to time-division and digital techniques, the same types of processors used to control analog SPC exchanges can still be used to control new-generation time-division switching. The increased use of digital techniques has indeed led to the increased use of computers in telephone exchanges. Not only are microprocessors now widely used in many of the subsystems from which exchanges are built up but general-purpose computers are increasingly used for network management, billing and administrative tasks directly associated with each exchange.

It is also of great potential importance that business computers are now beginning to be built which provide the same standards of reliability available at one time only in special telecommunications processors. New-generation business computers could well therefore be used to control tomorrow's telephone exchanges.

15 Signalling

In a telephony context, signalling means the passing of information and instructions from one point to another relevant to the setting up or supervision of a telephone call.

To initiate a call a telephone subscriber lifts the handset off its rest—in American English, "goes off hook". This off-hook state is a signal to the exchange to be ready to receive the number of the called subscriber. As soon as appropriate receiving equipment has been connected to the line, the exchange signals dial tone back to the calling subscriber who then dials the wanted number. On older exchanges, this information is passed via a rotary dial by a series of makes-and-breaks of the subscriber's loop, interrupting current flow. On more modern exchanges, voice-frequency musical tones are sent to the exchange as push buttons are pressed. These tones are usually called DTMF for Dual Tone Multi-Frequency, because each time a button is pressed two tones are sent out to line simultaneously, one from a set of four high frequencies, one from a set of four low frequencies. The subscriber in due course then receives advice from the exchange about the status of the call, either a ringing signal (indicating that the wanted line is being rung), an engaged or busy tone signal (indicating that the wanted line is already busy on another call), an equipment busy tone signal (indicating congestion somewhere between the called exchange and the calling line), or some other specialised tone.

These are the signals and tones with which telephone subscribers themselves are concerned. Telephone signalling is however also concerned with the signalling of information between exchanges.

Until recently all such signalling was carried on, or directly associated with, the same speech path as was to be used for the call being established or supervised. Various terms are commonly used in connection with these speech-path-associated signalling systems:

a) MF—multi-frequency, i.e. using voice-frequency tones.

b) MFC—multi-frequency compelled: this type of signal continues until the distant end acknowledges receipt and compels it to stop.

c) 1VF—one voice frequency: a single tone, sometimes pulsed in step with rotary dial impulses. 2600 Hz is a common 1VF tone.

d) 2VF—two voice frequencies: two tones, sometimes used together, sometimes separately.

e) Inband—a tone actually on the voice circuit itself, audible to anyone using the circuit (and so cannot be used during conversation).

f) Outband—signals directly associated with a voice circuit but either carried on separate wires or using a different frequency, just outside the commercial speech band of 300–3400 Hz. A frequency of 3825 Hz is often used for outband signalling.

All of these signalling systems have a number of limitations:

a) Relatively slow.

b) Limited information capacity.

c) Limited capability of conveying information which is not directly call-related.

d) Inability of some systems to send detailed information back to the calling end.

e) Inability of some systems to provide sufficient information for accurate itemised call billing.

f) Systems tend to be designed for specific application conditions.

g) Systems tend to be expensive because each circuit has to be equipped independently; there are no sharing techniques and no economies of scale.

The increased used of computer-controlled (or SPC) exchanges has led to the introduction of a completely different signalling concept. Instead of signalling being carried out on or directly associated with the voice channel carrying the conversation, there is now a move towards signalling being concentrated onto fast data circuits between the processors of the SPC exchanges concerned, leaving the voice circuits purely to carry voice signals. Signalling for several hundred long-distance circuits can be carried by a single fast data system, and substantial economies result.

A signalling system of this type has now been standardised by the body responsible for drawing up specifications for international use; this is called CCITT Signalling System No. 7. (CCITT means the International Consultative Committee for Telephony and Telegraphy.) No. 7 signalling has not only been designed to control the setting-up and supervision of telephone calls but of non-voice services also, such as word processors, teletex machines, etc. With common channel signalling systems such as CCITT No. 7, signalling is performed in both directions, with one signalling channel in each direction. This type of signalling has several attractive features:

Fig. 15.1 Common channel signalling (such as CCITT No. 7)

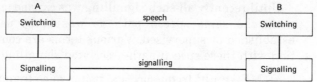

(a) *Common-channel signalling between A and B on an associated basis*

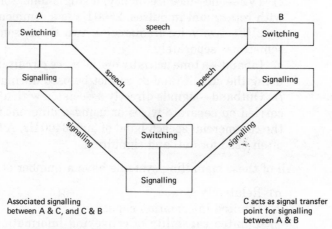

Associated signalling between A & C, and C & B

C acts as signal transfer point for signalling between A & B

(b) *Common-channel signalling between A and B on a quasi-associated basis*

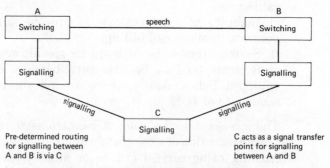

Pre-determined routing for signalling between A and B is via C

C acts as a signal transfer point for signalling between A and B

(c) *Common-channel signalling between A and B on a dissociated basis i.e. signalling can be routed and switched quite independently from the speech circuits being controlled a)* Signalling is completely separate from switching and speech transmission, and thus may evolve without the constraints normally associated with such factors.

b) Significantly faster than voice-band signalling.

c) Potential for a large number of signals.

d) Freedom to handle signals during speech.

e) Flexibility to change or add signals.

f) Potential for services such as network management, network maintenance, centralized call accounting.

g) Particularly economic for large speech circuit groups.

h) Economic also for smaller speech circuit groups due to the quasi-associated and disassociated signalling capabilities (see fig. 15.1).

i) Systems have been standardized for international use.

j) Can be used to control the setting-up and supervision of non-voice services and so will be important for ISDN (see page 124).

A common channel signalling system providing the signalling for many speech circuits must have a much greater dependability than on-speech-path signalling since random errors could disturb a large number of speech circuits. For this reason, provision must be made to detect and correct errors. Additionally, automatic re-routing of signalling traffic to a good back-up facility must occur in a situation of excessive error rate or on failure of a signalling link.

---

16 PBXs, PABXs, Keyphones, Centrex

When telephones were new and rare, very few companies had more than one line. The phone itself was carefully guarded and used only by The Boss. As time went on, the convenience of being able to talk to others without leaving your own desk encouraged companies to install phones for more and more of their staff—but here the first difficulty arose: cost. Each phone had to be complete with its own pair of wires all the way back to the central office/exchange, rented from the Telephone Company, and in many cases there was a charge for each call even when The Boss spoke only to an assistant in the next room.

There are basically two ways of tackling this problem:

a) Provide an "isolated" local telephone system in the office, an "intercom" with no connection at all to the public network. An automatic version of this is usually called a Private Automatic Exchange, PAX. Only a few senior people then have phones on outside lines.

b) Provide a switching system which enables the same telephone to be used for both internal and external calls, a Private Branch Exchange or PBX. Since the telephones themselves are not directly connected to an exchange/central office, it is usual for all the phones on such installations to be called "extensions". Early PBXs were looked after by operators; typically one girl could readily control a board serving up to about 50 extensions and 10 exchange lines. The operator dealt with all calls, internal and external, answered all incoming calls, and dialled out all the outgoing calls. Boards like this were called PMBXs, Private Manual Branch Exchanges.

This was fine; the operator knew where everyone was and it provided a built-in security service which stopped the office junior making expensive long-distance calls to his relations in Kansas City—the early and completely non-automatic ancestor of today's complex Telephone Information Management Systems (TIMS). But sometimes, operators being human, unacceptable delays occurred: if the operator simply had to leave her board for a while, neither outgoing nor incoming calls could be connected.

The development of automatic switching equipment soon led to the situation in which each instrument was able to control its own outgoing calls— it being about as easy to dial a number as to repeat it verbally to an operator—but even with these Private Branch Automatic Exchanges (PABXs), incoming calls depended on being answered by the operator. There is no easy way an ordinary outside call can be steered through a PABX to a particular extension.

Many modern public telephone central offices/exchanges apply a "time-out" to calls which are not answered within a specified period, usually one minute. This releases all the switches in order to stop the whole network becoming congested by connections which carry no traffic—and earn no money for the telephone company. If PABX operators are very busy (or if some of them are on unexpected sick leave), then by the time they get round to answering the last of the calling lines there is sometimes no-one there. It is not always the caller who has hung up in disgust, it is sometimes the switching equipment which has "timed-out" the call. This is probably one of the most frustrating aspects of trying to call busy companies: inadequate provision of operators must lose the companies concerned a lot of goodwill, and a lot of business.

One of the ways in which this particular difficulty can sometimes be overcome with modern switching equipment is to use central offices/exchanges with "in-dialling" or "dialling-in" lines. These are specially designed to enable outside callers to steer their own calls through the destination PABX. Another bright idea (which does not necessitate any changes in the central office or main exchange) is the use by the customer not of a centralized local switching system (a PABX) but of a system with very little centralized brains. Intelligence is distributed out to the telephones themselves; they have to be able to do more than just dial out digits. This was the origin of the keyphone system. Instead of incoming calls going only to the operator to answer, all the exchange lines were fed to all the extension instruments, giving all phones complete access to all lines. Any extension can answer incoming calls, all extensions can pick up outgoing lines and dial their own calls; no separate operator's position is needed, just an agreement by which nominated extensions answer calls and transfer them to the wanted extension.

This seems so simple that there has to be a snag, and there was: early versions of this sort of system needed large cables with a great many pairs of wires to each phone. If people's desks never had to be moved, this would not matter a great deal but in most active companies the reorganization of local layouts seems to be a frequent necessity as new departments or management structures are created, and desks have to be fitted into available space. Fortunately the chip came to the rescue here; it is now possible to purchase keyphone systems which do not need large cables, so it might be thought that it could now be a straight commercial fight between (a) systems using PABXs (i.e. a centralized switch) and (b) those in which much of the intelligence has been distributed to somewhat more complex keyphones. This is indeed the position today in many national administrations, the customer can choose either a or b.

More history now. When Common Control of central offices (mainly of crossbar switches) became economically possible in the 1950s, it was quickly realized that there was no purely technical reason for customers to have their own PABXs at all. Each of their extension phones can instead be served from the main switch or exchange; each is then given a standard number so incoming calls go straight to the required extension, with an automatic transfer to an assistance operator in the customer's premises if the extension fails to answer. Calls to the company's directory number also go straight to the assistance operator's console (often found at the receptionist's desk). The central office/exchange is programmed so that calls from one extension to another in the same company are ignored by the charging equipment, which would normally count each outgoing call as a unit call to be charged and paid for in due course. There is no PABX to purchase or to worry about or to maintain, it sounds too good to be true.

Cost is the main problem of course. A separate cable pair has to be rented from the telephone company linking each extension with the main central office/exchange and of course one complete line termination on the switch has to be rented also. Here a lot depends on the tariff policy being followed by the telephone company. If the central office/exchange is very close, the cable pairs may not be expensive. This service, using TelCo's central office as your own PABX, is usually called a Centrex service. There are variants of this available in some areas by which more of the switching equipment is located in the customer's premises; this reduces the number of cable pairs to be rented.

In some countries there is now yet another variant: Centrex-type service provided by a third party (e.g. the building landlord). When you rent an office in a block with this service, your office comes complete with mains power, water, sewage and telephone service: all the utility services are provided by the landlord. Ordinary PABXs are sometimes also installed in this way, one very large PABX in the basement serving each floor of a highrise building with different directory number groups available for the companies occupying each floor. Station message detail recording (SMDR) can be provided as part of all modern PABX installations. This enables the responsible manager to see where costs are being incurred and to take whatever action is appropriate. Telephone Information Management Systems (TIMS) have come a long way from their Kansas City origins!

Until about 1983 PABXs everywhere were analog devices, using space division switching. The adoption of digital technology (which in most countries began with large trunk and tandem switching centres like ESS4 and System X) has led quickly to the digitalization of PABXs. It is now possible in many areas to instal digital units which are not only cheaper than their analog counterparts but are able to switch both voice and non-voice signals. A single PABX can sometimes now be used as the controlling centre for a complete Local Area Network (LAN), able to establish calls between telephones, word processors, teletex machines and computers, all operating at different bit rates. Those multi-purpose switches have made complete Office Automation available now, not just a dream for the future.

So if you need telephones for a new office you must find out first exactly what is locally approved and available. In many areas it is now possible to get genuinely competitive bids for the provision of identical services by quite different methods. Most modern systems are able to provide service features which were expensive luxuries—or completely unavailable—only a dozen years ago.

17 Digital Exchanges

It has become conventional to divide telephone exchanges into two main categories:

1) *Space Division* (*or Analog*)—in which direct physical paths are established right through the exchange from one subscriber's line to another. Connections may be by metallic contacts (electro-mechanic step-by-step Strowger switches, rotary switches, crossbar switches or reed relays) or by solid-state analog devices.

2) *Time Division* (or *Digital*)—in which some or all of the switching stages in the exchange operate by shifting signals in time. Basically a connection is made between incoming and outgoing channels by transferring each PCM word (see section 33) from the time-slot of the incoming channel to that of the outgoing channel. This time-shifting is carried out by means of stores. Information is written into an address in a store, then during every cyclic scan the information from that particular address is read out so that it occupies the required outgoing time-slot.

Some purists object to the phrases "digital exchanges" and "digital switching"; these exchanges should, they say, really be called time-division. But market forces are such that the word "digital" is invariably used by manufacturers all over the world, even if many of the switching crosspoints in their exchanges are in fact pure space-division switches, often using reed relays.

The main reasons for going over to digital are

a) Lower first cost and lower annual charges than for analogue equipment. Maintenance staff savings are for example likely to be significant.

b) Space savings: digital exchanges in a fully digital environment will take up far less space than analog exchanges—in some circumstances less than 1/10th of the floor area will be needed.

c) Transmission improvement: the change from FDM to digital TDM transmission systems combined with the change from 2-wire space-division exchanges to digital exchanges (which are in effect all 4-wire devices) enables losses to be reduced significantly without having to invest heavily in new cable plant in local distribution networks.

d) The relative ease by which digital switching equipment can evolve to provide the many new services which customers are beginning to demand.

Most digital exchanges are built up from subsystems; the same subsystems can be put together to provide a variety of exchanges, for use at different points in the network (see figs 17.1 and 17.2).

The simplification of function which follows the final elimination of FDM and analog trunks and junctions will be apparent. When this stage is reached, exchanges will all take up a fraction of the floor area now occupied, with great consequent savings.

The function of the concentration stage is to interconnect subscriber lines with the main switching subsystem in the exchange, the digital or group switching stage. Out of every 1000 lines on an exchange, the probability is that not more than about 100 will be making calls at any one instant of time, even during busy periods. It would therefore be a waste of money to provide so much switching equipment that all the subscribers on the exchange would be able to talk at the same time. The concentration stage has to be carefully designed to provide the grade of service required by the administration, with the greatest possible economy. Business lines are used much more than residential lines, so it is

Fig. 17.1 Digital trunk exchange

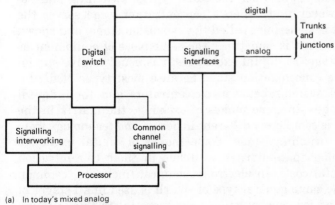

(a) In today's mixed analog and digital environment

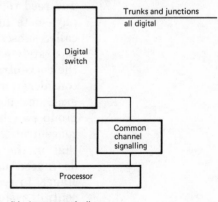

(b) In tomorrow's all-digital environment

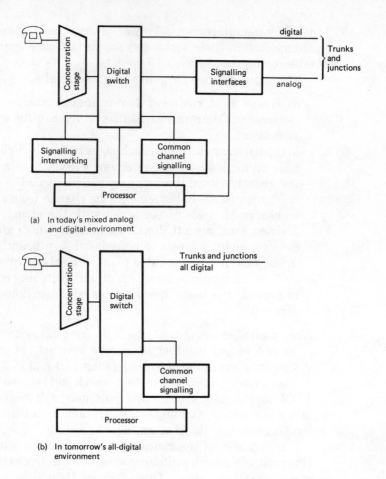

Fig. 17.2 Digital local exchange

(a) In today's mixed analog and digital environment

(b) In tomorrow's all-digital environment

not possible to concentrate business lines to the same extent as low-calling-rate residential lines. (Junctions to other exchanges are not concentrated at all; every trunk and junction circuit always has direct access to the group switch.)

This principle of concentration of subscriber's lines, followed by switching, followed by a reverse path through the concentrator out to the called subscriber, is not peculiar to digital exchanges. Step-by-step, crossbar and reed relay exchanges also use the same type of switching philosophy—note here however that, when a concentration stage serves the called subscriber, it is sometimes called the expansion stage, and shown separately even though it is sometimes the same piece of equipment as the concentration stage. In digital exchanges it is usual to give special consideration to the concentration stage because most types of digital exchange use PCM systems between concentrators and their digital group switching stages. In some makes of exchange the matrix in the concentration stage is completely different in design and technology from that in the group switching stage. Some makes of digital exchange do indeed use analog space-division switches in their concentration stages, to reduce total costs. In electro-mechanical (including common control) systems, the same general type of switch is used in all stages of switching: those used for concentration have more inlets than outlets;

those used for switching usually have the same number of inlets as outlets; those used for expansion stages have more outlets than inlets. (Since the same technology is used for all stages it is not usual to consider concentration stages as being anything other than part of the switching matrix in these earlier types of exchange.)

The interface with the subscriber's line is at the present time the most costly part of all digital exchanges, largely because a 10 000 line exchange has to have 10 000 of these units, each with all the features needed to interwork with various types of subscribers' lines. It is customary to describe these features as the "Borscht" functions, based on the initials of the key words:

Battery feed to line—there is normally no active power source at a subscriber's telephone; all the power (needed to drive the microphone and key pad) is fed out from the exchange along the subscriber's line.

Overvoltage protection—solid state devices are very sensitive to high voltages, and rapid-action protective devices have to be provided in each line circuit so that if lightning does happen to strike an external line the exchange will not be put out of action.

Ringing current injection and ring trip detection—the bell at the called subscriber's telephone has to be rung; this means that quite a high voltage a.c. signal (sometimes about 70 volts) has to be connected to the line, to ring the bell; and as soon as the handset has been picked up off its cradle (gone "off hook") the ringing must be tripped and disconnected.

Supervision of the line—equipment has to be provided which continually monitors the line so that as soon as it goes "off hook" (and a continuous d.c. path provided through the instrument) the exchange connection is activated and dial tone sent out to the caller. Dial pulses represent breaks in the continuous d.c. loop; these have to be detected and counted so that the exchange knows what number is required. When the caller finally clears down, by going back "on hook", the exchange must break down the established call and note the time at which this has been done, for charging purposes.

Codec (short for encoder plus decoder)—this turns the analog signal received from the telephone instrument into a digital signal ready to be multiplexed with others into a PCM system. Incoming signals are similarly decoded from digital to analog before being sent out to the subscriber's instrument. Some digital exchanges have one codec per line; some share codecs between several lines; some use even fewer codecs by placing them between their concentration stages and the links going to their group switching stages.

Hybrid for 2-wire to 4-wire conversion. The line to an ordinary telephone subscriber uses one pair of wires, for both directions of conversation; this is called a 2-wire circuit. The circuits inside a digital exchange use two electrically separate paths, one for each direction of transmission; this is called a 4-wire circuit. To join together a 2-wire and a 4-wire circuit, a special device called a hybrid coil is used; this allows speech from the 2-wire line to enter the 4-wire transmit path, and speech from the 4-wire receive path to pass to the 2-wire line, but it blocks speech incoming on the 4-wire receive path from going out again on the 4-wire transmit path of the same circuit. (Ordinary electro-mechanical exchanges use 2-wire circuits all the way through so there is no need for hybrids in these exchanges.)

Testing of both line and equipment—it is necessary to be able to test the subscriber's line electrically so that faults may be located and cleared.

Any form of time-division switching is necessarily a one-way function; to provide a single bi-directional speech circuit through an exchange, two channels therefore have to be switched through, one for each direction of transmission. Fig. 17.3 shows how this is done in some exchanges.

Fig. 17.3 Time-space-time switching

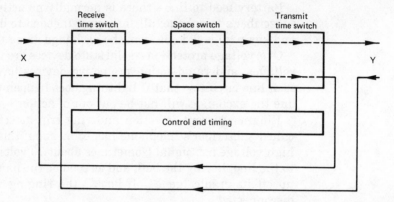

The use of common channel signalling systems such as CCITT No. 7 will enable a single exchange to be used to switch all types of digital information services; it will not be necessary to have separate exchanges for record services such as telex. (See section 35.)

Part D Cable, Radio and Transmission

18 Local Distribution Networks

The connection between a subscriber and the local telephone exchange consists of a pair of wires in a telephone cable. Since a large telephone exchange may have 10 000 or more subscribers, the local line network can be quite complicated, particularly because provision must be made for fluctuating demand. The local line network is provided on the basis of forecasts made of the future demand for telephone service, the object being to provide service on demand and as economically as possible. Since the demand fluctuates considerably there is the problem of forecasting requirements and deciding how much plant should be provided initially and how much at future dates. No matter how carefully the forecasting is carried out, some errors always occur and allowance for this must be made in the planning and provision of cable, i.e. the local line network must be flexible. A network must be laid out so that the situation should not arise where potential subscribers cannot be given service in some parts of the exchange area while in other parts spare cable pairs remain.

The modern way of laying out a local line network is shown in fig. 18.1. Each subscriber's telephone is connected to a distribution point, such as a terminal block on a pole or a wall. The distribution points are connected by small distribution or secondary cables to cabinets. Primary or main cables then connect these cabinets to the telephone exchange.

It is usual to provide secondary cables on the basis of an expected life of about 15 years and the much larger primary cables for only about 5 years. This does not mean that these cables are expected to be scrapped after these periods; they are expected by then to be fully utilised and to require supplementing by additional cables. Cabinets are sometimes called "flexibility points": if demand is much heavier than forecast in one part of the area served by the cabinet, and less in another part, cable pairs may be connected through at the cabinet to the faster developing area.

External plant usually represents the largest single component of the capital assets of a national telephone system so it is important that full consideration be given to this subject. All too often, however, local cabling is given only low management level attention despite its importance to subscribers and to the financial well-being of the administration.

Capital expenditure has to be minimized while providing sound engineering combined with flexibility.

From the outset, line planning specialists must liaise with transmission and switching planners to designate the fundamental local network parameters in accordance with CCITT recommendations, keeping in mind the economic advantages of apportioning the largest amount possible of reference equivalent to the local distribution network. The finer the gauge cable that can safely be used in the network, the lower the costs will be.

Using this information together with the forecast demand, the practical wire centres for each exchange location can be determined—using a computer programme if necessary. Economic studies should be carried out if the proposed site for a new exchange proves to be a long way away from the calculated wire centre. Types of construction to be used, sizes of cables and conductor gauges must then be chosen, taking into account the geological and topographical conditions in both urban and rural areas.

It is usually economically desirable to use underground ducted cable from exchange to flexibility points (cabinets) and then either go underground or overhead to distribution points depending on circumstances (e.g. in some countries it is possible to arrange for joint use of poles with the local electricity supply authority). A ducted system, although initially more costly than a direct buried system, provides much greater flexibility for the future installation of additional cables, and facilitates cable repairs or replacement.

At the present time, polyethylene-insulated conductors and polyethylene-sheathed cable incorporating an aluminium screen and water vapour barrier is the most economical type of cable to be used in ducts. Main or primary cables from exchanges to cabinets should wherever possible be airspaced with protection provided by continuous-flow dry-air pressurization. Secondary or distribution cables from cabinets to distribution points (and smaller main cables) should be jelly-filled, not pressurized. Aerial cables should be of similar design but with an in-built self-supporting catenary wire in the figure-of-eight mode.

Many modern telephone switching systems incorporate concentration stages which can be either co-located (i.e. in the same building as the rest of the exchange) or remote (i.e. many kilometres away, fed by PCM systems back to the main part of the exchange). The use of remote concentrators to serve telephone subscribers in small towns means that cable pairs to such subscribers will not in future need to be of the heavy gauge which was in the past needed to feed them all the way back to the nearest city; comparatively fine gauge (and therefore cheaper) cables can now be used for distribution to these subscribers. (See fig. 18.2 and fig. 18.3.)

The economic reasoning behind the use of remote concentrators may be summarized:

a) The transmission losses which can be accepted in the circuits between subscribers and exchange can be largely allocated to the secondary cable network (between concentrator and distribution point serving the subscriber) because the circuits between concentrator and exchange are

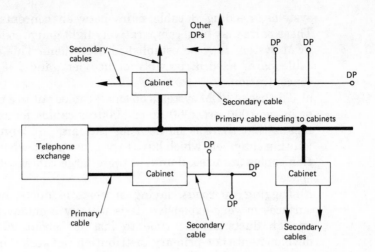

Fig. 18.1 Layout of a telephone exchange area

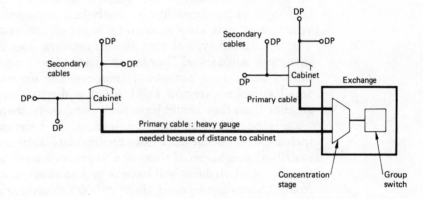

Fig. 18.2 Traditional-type cable distribution

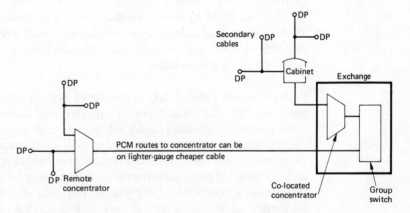

Fig. 18.3 Use of a remote concentration stage

low-loss channels in PCM systems. This means that a lighter gauge (and therefore cheaper to purchase, handle and lay) cable may be used in this extensive part of the distribution network.

b) Most concentrators are designed to provide an efficient concentration ratio of about 10 to 1, i.e. for every 1000 subscribers lines connected, about 100 PCM channels are needed in service between concentrator and exchange. 100 channels may be provided by only four 30-channel PCM systems, needing 8 cable pairs between concentrator and exchange. These pairs may be of comparatively light gauge so long as the 2 Mbit/sec PCM system may be established over them. This compares with 1000 cable pairs needed, possibly of heavier gauge, if there is no remote concentration.

c) The use of PCM systems on small cables out to a concentrator (instead of using 1000 pr, 2000 pr or 3000 pr cables for primary distribution) means that there is no need for the large underground manholes and jointing chambers which have to be constructed when very large distribution cables are used. Joints in these large cables take up a great deal of space.

d) Digging up roads, laying of nests of ducts, and reinstating road surfaces is very expensive. It is usually economically desirable to lay enough ducts at one time to last for about 20 year's growth and development. For primary distribution networks, in which most cables are so large that each one occupies a complete duct on its own, this means that most of the ducts in any newly-laid nests will be empty for many years—duct nest sizes have to be based on forecasts of future demand. The traditional type of distribution network does necessarily therefore involve a substantial "burden of spare plant", i.e. most of the newly-buried plant in a network is non-revenue earning. With the use of smaller cables carrying PCM, all new duct nests can be made much smaller than they would have been—and more than one main cable can usually be pulled in to a single duct. In practical terms a single 4 inch/10 cm diameter duct can accommodate only one 2000 pair primary distribution cable so, if there is a forecast demand for 20 000 lines in an area, at least 10 ducts will have to be laid to serve it. If PCM and remote concentrators can be used, these 20 000 subscribers could be fed satisfactorily by about 2000 channels. These could be provided by only 70 30-channel PCM systems, needing only 140 cable pairs to carry them. With the special cables now available which are able to carry PCM systems on every pair without mutual interference, these 140 pairs would not even fill a single duct.

It is of course not entirely a one-sided argument; remote concentrators need accommodation and power supplies, and have to be maintained. As time goes on it seems probable however that digital concentrators will become more and more compact and will soon usually be accommodated in roadside cabinets like the cable cross-connection cabinets now in general use. Or perhaps concentrators may soon be designed round so few integrated circuits that they will be put into sealed canisters which can be jointed in to the cable network and installed inside ordinary underground jointing chambers and manholes.

In the past, most telecoms administrations all over the world have regarded local line planning as a function which can safely be carried out separately from other system planning duties. Now that it is becoming economically attractive for the "front end" of each exchange to be located close to the subscribers it serves, it is no longer possible to treat such planning in isolation.

19 Carrier Working: Groups and Supergroups

In section 9 it was shown how a speech signal can, by amplitude modulation, be changed in frequency from its original audio frequency (of 300–3400 Hz) up to a higher "carrier" frequency. Most of the world's long-distance telephony systems utilise 12-channel groups: twelve voice channels are all changed in frequency in this way so that a complete group of 12 channels occupies a 48 kHz bandwidth, basically from 60–108 kHz.

Fig. 19.1 is a block schematic of the transmitting equipment required for channels 1 and 2 of a standard 12-channel group. The audio input signal to a channel is applied to a balanced modulator together with the carrier frequency appropriate to that channel. The input attenuator ensures that the carrier voltage is 14 dB higher than the modulating signal voltage, as required for correct operation of the modulator. The output of the modulator consists of the upper and lower sideband products of the amplitude modulation process together with a number of unwanted components.

Fig. 19.1 Schematic diagram of transmitting equipment for a carrier system

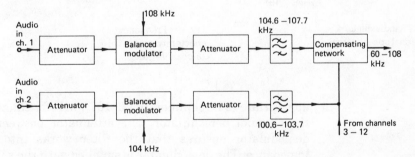

Following the modulator is another attenuator whose purpose is twofold: firstly, it ensures that the following band-pass filter is fed from a constant-impedance source—a necessary condition for optimum filter performance—and secondly, it enables the channel output level to be adjusted to the same value as that of each of the other channels. The filter selects the lower sideband component of the modulator output, suppressing all other components. To obtain the required selectivity, channel filters utilizing piezoelectric crystals are employed. The outputs of all the twelve channels are combined and fed to the output terminals of the group. The transmitted bandwidth is 60.6–107.7 kHz, or approximately 60–108 kHz.

The equipment appropriate to channels 1 and 2 at the receiving end of the 12-channel group is shown in fig. 19.2. The composite signal received from the line, occupying the band 60–108 kHz, is applied to the twelve, paralleled, channel filters. Each filter selects the band of frequencies

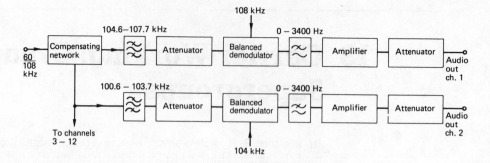

Fig. 19.2 Receiving equipment for a carrier system

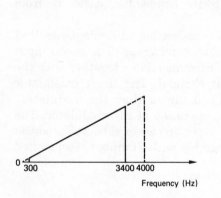

Fig. 19.3 Bandwidth for a commercial speech circuit

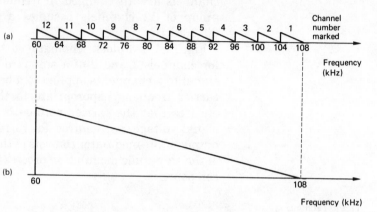

Fig. 19.4 Frequency spectrum diagrams of a 12-channel group

appropriate to its channel, 104.6–107.7 kHz for channel 1, and passes it to the channel demodulator. The attenuator between the filter and the demodulator ensures that the filter works into a load of constant impedance. The demodulator is supplied with the same carrier frequency as that suppressed in the transmitting equipment. The lower sideband output of the demodulator is the required audio-frequency band of 300–3400 Hz and is selected by the low-pass filter. The audio signal is then amplified and its level adjusted by means of the output attenuator.

The assembly of the basic 12-channel carrier group can be illustrated by means of a frequency spectrum diagram. The spectrum diagram of a single channel is given in fig. 19.3; the actual speech bandwidth provided is 300–3400 Hz but a 0–4000 Hz bandwidth must be allocated per channel to allow a 900 Hz spacing between each channel for filter selectivity to build up. Fig. 19.4*a* shows the frequency spectrum diagram for the 12 channels forming a group; the carrier frequency of each channel is given and so are the maximum and the minimum frequencies transmitted. It can be seen that all the channels are inverted; that is, the lowest frequency in each channel corresponds to the highest frequency in its associated audio channel, and vice versa. Since all the channels are inverted, the group may be represented by a single triangle as shown by fig. 19.4b.

The 12-channel system can be used as a building block for the next larger assembly stage or as a system which can be transmitted to line in its own right.

Five 12-channel groups can be combined to form a 60-channel supergroup, and five supergroups make up a 300-channel mastergroup. Three mastergroups then make up a 3872 kHz bandwidth supermaster group. Alternatively, 15 supergroups may be assembled direct to form a hypergroup, sometimes called a 15 supergroup assembly.

20 Submarine Cables

Submarine telecommunication cables have been with us for many years; the first such submarine cable of any significant length was laid in 1850, and the first specialized cable-laying ship was launched in 1872. But early cables were all for telegraphy; the distortionless wider band needed for telephony could not economically be transmitted over long lengths of cable until electron tube and component designers had produced long-life trouble-free units, capable of operating for many years in the depths of the oceans without fault incidence.

These units (called repeaters) were designed to amplify incoming signals to offset the transmission losses incurred in the previous section of cable. It was also necessary to provide equalization to offset the way different frequencies in the wideband information signal had been attenuated by different amounts. Such items were produced in the USA in the 1950s, and repeaters suitable for insertion in deep-sea cables were soon designed in America, Britain, France and Germany.

The first major submarine telephone cable was TAT-1, laid across the Atlantic from Scotland to Newfoundland in 1956. It had a capacity of 50 voice circuits. (It is no longer in use.) With the advance of technology it has become possible for wider bandwidths, meaning more circuits, to be transmitted: the latest transatlantic cable, TAT-8, to be brought into service in 1988, has an effective capacity of 40 000 voice circuits.

A submarine cable is sometimes under considerable tension, especially when it is being picked up from the sea bed, so great tensile strength is necessary in addition to the ability of cable and repeaters to withstand the high pressures of deep waters. Early submarine telephone cables were made up in a generally similar way to 19th century submarine telegraph cables, with one or more layers of heavy steel armour wire to provide the necessary protection (fig. 20.1). The cores of the two types of cable were of course quite different; telegraph cables usually had a heavy, well-insulated central copper conductor to carry the signal current, with the return current normally flowing through the sea itself, whereas wide-band telephone cables are of coaxial type, i.e. they have a central copper conductor, surrounded by a carefully dimensioned insulant, then copper tapes as a return-path outer conductor, then more insulant, then the outer armour wires (fig. 20.2).

In the 1960s there was a breakthrough in submarine cable design for use in deep waters (e.g. deeper than about 400 fathoms): instead of cables being made up with steel wire armouring on the outside of the cable, a comparatively lightweight high tensile steel rope was put at the centre of the cable, with a circumferential strip of copper compacted on to the steel rope to act as the central conductor, with a layer of polyethylene insulant, then a layer of aluminum or copper as the return conductor, then more polyethylene, on the outside for protection (fig. 20.3). In shallow waters it is still considered safer to continue to use conventionally armoured cables, but the introduction of this lightweight design for the long deep sea sections of cross-ocean routes has given submarine cable systems a new lease of life. Cable systems are now used together with satellite radio systems. There is an economic and security need for both types of communications media to be used; most countries feel that it would be unwise to depend completely on any one system, and no doubt within a few years fibre optics will enable cable manufacturers to provide circuits even more economically than is now possible.

The lists which follow give brief descriptions of all the modern Transatlantic and Transpacific submarine telecommunications cables:

Submarine Cable Systems across the North Atlantic

TAT-1 The first transatlantic telephone cable. Twin cables from Oban, Scotland to Clarenville, Newfoundland, then single cable to Sydney Mines, Nova Scotia, Canada. Total length 4210 cable miles, 118 repeaters, nominal capacity 50 voice circuits. In service in 1956, retired in 1978.

TAT-2 Twin cables from Penmarch, France to Clarenville, Newfoundland. Total length 4418 cable miles, 114 repeaters, nominal capacity 48 voice circuits. In service 1959.

TAT-3 Single cable from Widemouth Bay, England to Tuckerton, USA. Total length 3518 cable miles, 182 repeaters, nominal capacity 138 voice circuits. In service 1963.

TAT-4 Single cable from St Hilaire de Riez, France to Tuckerton, USA. Total length 3599 cable miles, 186 repeaters, nominal capacity 138 voice circuits. In service 1965.

TAT-5 Single cable from Conil, Spain to Green Hill, USA. Total length 3461 cable miles, 361 repeaters, nominal capacity 845 voice circuits. In service 1970.

TAT-6 Single cable from St Hilaire de Riez, France to Green Hill, USA. Total length 3396 cable miles, 694 repeaters, nominal capacity 4000 voice circuits. In service 1976.

TAT-7 A 4000 circuit cable between USA and UK. In service 1983.

TAT-8 The first optic fibre transatlantic cable. Initial capacity 8000 voice circuits (1988) but capable of expansion to 40 000 circuits.

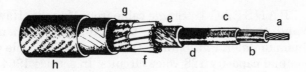

Fig. 20.1 Telegraph cable
a copper conductor
b gutta percha insulation
c brass tape
d bituminous wax tape
e jute
f wire armouring
g compounded tape
h jute servings with
   protective compound

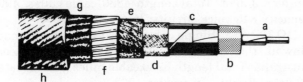

Fig. 20.2 Coaxial telephone
cable
a central copper
   conductor
b polythene insulation
c return copper conductor
d impregnated fabric tape
e jute serving
d armouring wires
g impregnated fabric tape
   on each wire
h jute outer serving

Fig. 20.3 Lightweight
coaxial telephone cable
a composite high tensile
   steel stress member
b central copper
   conductor
c polythene insulation
d aluminium return
   conductor tape
e polythene film separator
f aluminium screening
   tapes with polythene film
   interleaved
g impregnated protective
   cotton tape
h polythene sheath

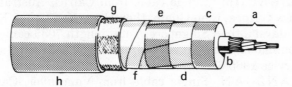

PTAT-1 (1988) and PTAT-2 (1992)   Cables similar to TAT-8 but for use by the private sector, to be leased in 45 Mbit/s segments.

SCOT-ICE plus ICE-CAN   A single cable, 24 voice circuit system installed in 1962, between Gairloch (Scotland), the Faeroes, Iceland, Greenland and Corner Brook, Newfoundland. Total length 2462 cable miles, 104 repeaters.

CANTAT-1   Single cable from Oban, Scotland to Corner Brook, Newfoundland. Total length 2073 cable miles, 90 repeaters, nominal capacity 80 voice circuits. In service 1961.

CANTAT-2   Single cable from Widemouth Bay, England to Beaver Harbour, Canada. Total length 2805 cable miles, 473 repeaters, nominal capacity 1840 voice circuits. In service 1974.

CANTAT-3   A possible 5000 voice circuit cable being discussed for provision by 1986, details not yet finalized.

COLOMBUS   Single cable from Canary Is., Spain to Camuri, Venezuela. Total length 3239 cable miles, 503 repeaters, nominal capacity 1840 voice circuits. In service 1977.

BRACAN-1   Single cable from Canary Is., Spain to Recife, Brazil. Total length 2649 cable miles, 138 repeaters, nominal capacity 160 voice circuits. In service 1973.

Submarine Cable Systems across the Pacific
HAW-1   Twin cable from Point Arena, California, USA to Hanauma Bay, Hawaii. Total length 2210(×2) cable miles, 114 repeaters, nominal capacity 51 voice circuits. In service 1957.

HAW-2   Single cable from San Luis Obispo, California, USA to Makaha, Hawaii. Total length 2383 cable miles, 123 repeaters, nominal capacity 142 voice circuits. In service 1964.

HAW-3   Single cable from San Luis Obispo, California, USA to Makaha, Hawaii. Total length 2379 cable miles, 248 repeaters, nominal capacity 845 voice circuits. In service 1974.

TRANSPAC-1   Single cable from Makaha, Hawaii to Midway, then Wake, then Agana, Guam, then Ninomiya, Japan, with a spur from Guam to Baler, Philippines. Total length 6750 cable miles, 356 repeaters, nominal capacity 128 voice circuits. In service 1964.

TRANSPAC-2   Single cable from Makaha, Hawaii to Guam, then Chinen, Japan. Total length 4880 cable miles, 490 repeaters, nominal capacity 845 voice circuits. In service 1975.

COMPAC   Single cable from Vancouver, Canada to Keawaula Bay, Hawaii, then Suva, Fiji, then Auckland, New Zealand, then Sydney, Australia. Total length 8230 cable miles, 322 repeaters, nominal capacity 80 voice circuits. In service 1963.

SEACOM   Single cable from Cairns, Australia to Madang, Papua New Guinea, then Guam, then Hong Kong, then Kota Kinabalu, Sabah, Malaysia, then Singapore. Total length 7085 cable miles, 353 repeaters, nominal capacity 80 voice circuits (160 between Guam and Australia). In service 1965/67.

ANZCAN   Single cable from Vancouver, Canada to Hawaii, Fiji, Norfolk Island and Sydney, Australia, with spur from Norfolk to Auckland, New Zealand. Total length 15 000 km (8000 cable miles) with over 1000 repeaters and capacity of 1380 voice circuits. In service 1984.

21 Optic Fibres

The use of a light wave as carrier, to be modulated by an information signal in the same way as these signals can modulate radio waves, was for many years considered impracticable. The reason for this was that light was emitted as a random series of energy pulses, generated largely by accelerating electrons and by electrons changing their energy levels inside atoms. Even light from what seemed to be one-colour sources like a sodium vapour lamp was found not to be a single simple sinusoidal wave but a whole series of waves, all differing in phase (and sometimes also in frequency) from each other.

The laser (light amplification by the stimulated emission of radiation) and the LED (light-emitting diode) were the new technology inventions which produced an answer to this problem. They are both devices which emit optical radiation as a direct result of applied voltages and electron movement. Both can be pulsed on and off very rapidly, and some lasers can also produce coherent light, i.e. light of a single frequency with all the waves in phase.

Although many lasers and LEDs are able to produce outputs in the visible light band, most current optic fibre telecommunications systems use signals of wavelengths 0.8 μm or 1.3 μm, both in the infra-red band. These are still called optical systems: even though the signals cannot be seen, they are transmitted in exactly the same way as visible light signals. The main reason why engineers wanted to be able to modulate coherent light was to take advantage of the tremendous bandwidths which could be carried by these very high frequencies; the figures below are typical figures, indicating the orders of magnitude involved.

	Carrier Wave		Possible bandwidth per system
	Frequency	Wavelength	
HF radio	3 MHz	30 m	16 kHz (4 voice channels)
Microwave radio	6 GHz	5 cm (10^{-2}m)	4 MHz (960 voice channels)
Optic fibre	100 000 GHz	1 μm (10^{-6}m)	Several thousand MHz (Hundreds of thousands of voice channels—but only a few thousand are possible with current technology)

An optical fibre cable consists of a glass core that is completely surrounded by a glass cladding. The core performs the function of transmitting the light waves, while the purpose of the cladding is to minimize surface losses and to guide the light waves. The glass used for both the core and the cladding must be of very high purity since any impurities present will cause some scattering of light to occur. Two types of glass are commonly employed: silica-based glass (silica with some added oxide) and multi-component glass (e.g. sodium borosilicate). (Some new optic fibres do not use glass at all, but special types of plastic; these are usually cheaper to make than very pure glass but introduce greater attenuation.) Fibres now being manufactured are so free from impurities that very little energy need be lost as the signals travels along—an attenuation of less than 1 dB per kilometre is not uncommon for the latest high-purity silicons.

A major constraint with optic fibres (apart from the straightforward one of attenuation) is that, since the wavelength of light is very short, a light wave signal injected into one end of an optic fibre (sometimes called an optic wave-guide) does not merely travel straight down the middle of the core. It swings from side to side, continually being reflected or refracted back from the core/cladding surface. Clearly a pulse going straight down the middle will reach the end just before those parts of the same pulse signal which have zigzagged along, taking a longer path. This places a restriction on the maximum possible bit rate that may be transmitted satisfactorily in the fibre.

It would have been ideal if right from the beginning we could have used fibres made with tiny diameters that were comparable with the wavelength of the optical signal being used, so that no zigzagging was able to take place, but the manufacture and jointing of such high-precision fibres still presents considerable difficulty. Today's more common thicker-core fibres are called multimode (because many different modes of transmission are possible). Fibres with very small diameter cores are called monomode (because only a single mode of transmission is possible). There are, therefore, three basic types of optical fibre:

1) *Stepped-index multimode* The basic construction of a stepped-index multimode optical fibre is shown in fig. 21.1a and its refractive index profile is shown by fig. 21.1b. It is clear that an abrupt change in the refractive index of the fibre occurs at the core/cladding boundary. The core diameter $2r_1$ is usually some 50–60 μm but in some cases may be up to about 200 μm. The diameter $2r_2$ of the cladding is standardized, whenever possible, at 125 μm.

Stepped-index multimode fibre produces large transit time dispersion (fig. 21.2), so its use is restricted to applications such as those involving comparatively low-speed data signals.

2) *Stepped-index monomode* Fig. 21.3a shows the basic construction of a stepped-index monomode optical fibre and fig. 21.3b shows its refractive index profile. Once again the change in the refractive indices of the core and the cladding is an abrupt one but now the dimensions of the core are much smaller. The diameter of the core is of the same order of magnitude as the wavelength of the light to be propagated; it is therefore in the range 1–10 μm. The cladding diameter is the standardized figure of 125 μm.

Stepped-index monomode fibres are at present difficult and expensive to manufacture and to join, so most of the stepped-index optic-fibre telecommunications systems now installed use multimode fibres. However, technological difficulties are being overcome and the use of monomode fibres is increasing throughout the world.

3) *Graded-index multimode* The basic construction of a graded-index multimode optical fibre is the same as that of the stepped-index multimode fibre shown in fig. 21.1a. The core diameter is also in the range 50–60 μm and the cladding diameter 125 μm. The refractive index of the inner region or core is highest at the centre and then decreases parabolically towards the edges (fig. 21.5) to that of the cladding material. This means that light waves will be refracted back from the outer boundary of the fibre, not reflected as with stepped-index fibres (see fig. 21.6). So waves will go straight down the centre of the core, or zig-zag from side to side as they do in stepped-index fibres but in a "smoother" manner. The main difference is, however, that waves which zig-zag along in a graded-index fibre pass through regions with a lower refractive index than the central part of the core, so, although they travel a greater distance, it is at a higher velocity. The effect of this is to reduce the differences in the times taken by the many different modes; ideally, all modes then arrive at the distant end in exact synchronism.

No matter which of the three possible types of propagation is used, the dimensions of the outer medium or cladding must be at least several wavelengths. Otherwise some light energy will be able to escape from the system, and extra losses will be caused by any light scattering and/or absorbing objects in the vicinity.

Fig. 21.1 (a) Stepped-index
multimode optical fibre;
(b) refractive index profile

(a)

(b)

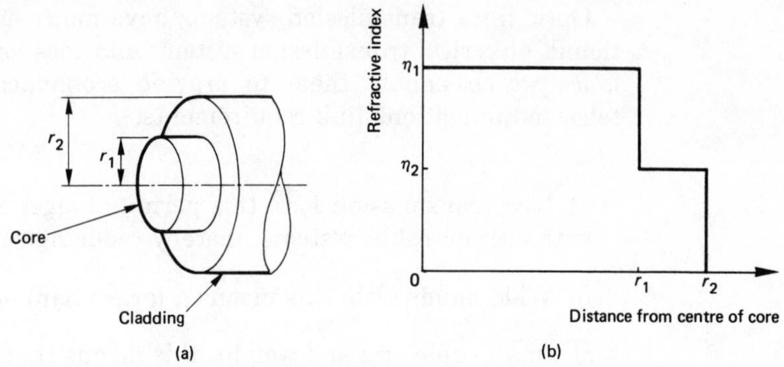

Fig. 21.2 Multimode
propagation in a stepped-
index fibre

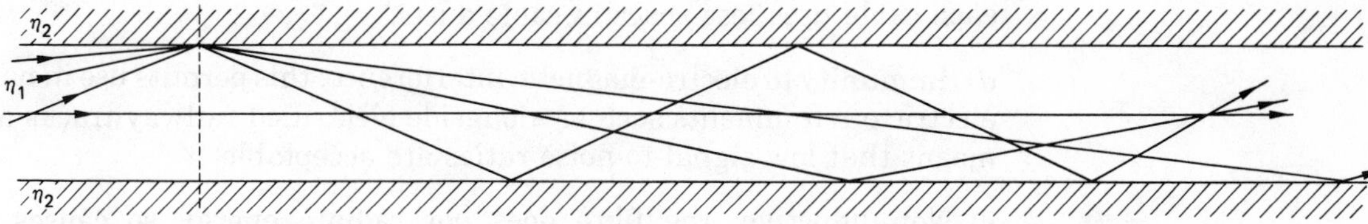

Fig. 21.3 (a) Stepped-index
monomode optical fibre;
(b) refractive index profile

(a)

(b)

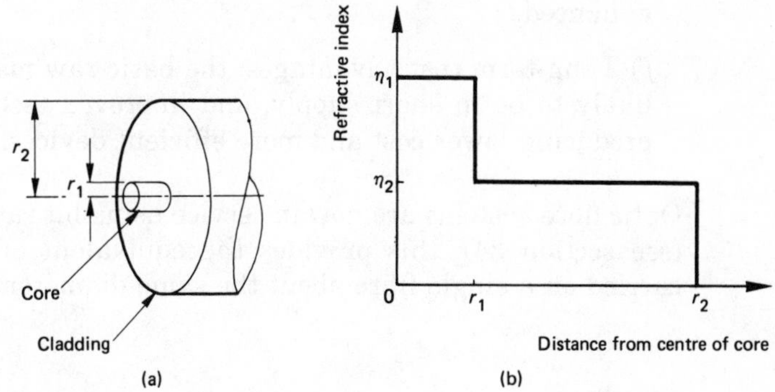

Fig. 21.4 Monomode
propagation in a stepped-
index fibre

Fig. 21.6 Multimode
propagation in a graded-
index fibre

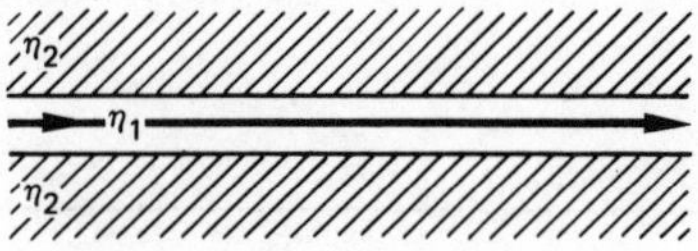

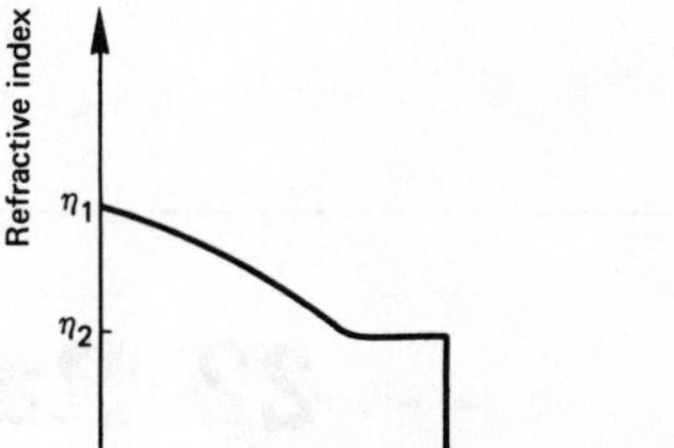

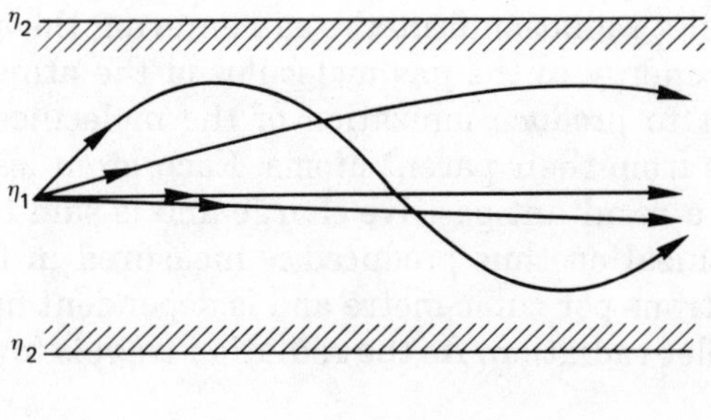

Fig. 21.5 Refractive index
profile of grade-index
multimode optical fibre

Optic fibre transmission systems have many advantages over "traditional" electrical transmission systems and possess significant characteristics which enable them to provide economical solutions to many telecommunications link requirements:

a) Low transmission loss: this permits longer repeater sections than with coaxial cable systems, thereby reducing costs.

b) Wide bandwidth: this means a large channel carrying capacity.

c) Small cable size and weight: this means that drums of cable can be handled economically (by people instead of by fork lift trucks) and that each cable uses less space in cable ducts.

d) Immunity to electro-magnetic interference: this permits use in noisy electric environments such as alongside electrified railway tracks and means that low signal-to-noise ratios are acceptable.

e) Non-inductive: the fibre does not radiate energy so causes no interference to other circuits. Communications security is thereby enhanced.

f) Long-term cost advantages: the basic raw material, silica, is never likely to be in short supply, and improved technology is continually producing lower cost and more efficient devices.

Optic fibre systems are now in service using bit rates of up to 560 Mbit/s (see section 34); this provides the equivalent of 7680 voice channels carried on a single fibre about the same diameter as a human hair.

22 Radio Propagation

1 *The Ionosphere*

Ultra-violet radiation from the sun entering the atmosphere of the earth supplies energy to the gas molecules of the atmosphere. This energy is sufficient to produce ionization of the molecules, that is remove some electrons from their parent atoms. Each atom losing an electron in this way has a resultant positive charge and is said to be ionized.

The ionization thus produced is measured in terms of the number of free electrons per cubic metre and is dependent upon the intensity of the ultra-violet radiation. As the radiation travels towards the earth, energy

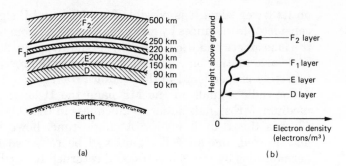

Fig. 22.1 Layers in the ionosphere

(a)

(b)

is continually extracted from it and so its intensity is progressively reduced. The liberated electrons are free to wander at random in the atmosphere and in so doing may well come close enough to a positive ion to be attracted to it. When this happens, the free electron and the ion recombine to form a neutral atom. Thus a continuous process of ionization and recombination takes place.

At high altitudes, the atmosphere is rare and little ionization takes place. Nearer the earth the number of gas molecules per cubic metre is much greater and large numbers of atoms are ionized; but the air is still sufficiently rare to keep the probability of recombination at a low figure. Nearer still to the earth, the number of free electrons produced per cubic metre falls, because the intensity of the ultra-violet radiation has been greatly reduced during its passage through the upper atmosphere. Also, since the atmosphere is relatively dense, the probability of recombination is fairly high. The density of free electrons is therefore small immediately above the surface of the earth, rises at high altitudes, and then falls again at still greater heights. The earth is thus surrounded by a wide belt of ionized gases, known as the ionosphere.

In the ionosphere, layers exist within which the free electron density is greater than at heights immediately above or below the layer. Four layers exist in the daytime (the D, E, F_1 and F_2 layers) at the heights shown in fig. 22.1.

The heights of the ionospheric layers are not constant but vary both daily and seasonally as the intensity of the sun's radiation fluctuates. The electron density in the D layer is small when compared with the other layers. At night-time when the ultra-violet radiation ceases, no more free electrons are produced and the D layer disappears because of the high rate of recombination at the lower altitudes. The E layer is at a height of about 100 km and so the rate of recombination is smaller. Because of this, the E layer, although becoming weaker, does not normally disappear at night-time. In the daytime, the F_1 layer is at a more or less constant height of 200–220 km above ground but the height of the F_2 layer varies considerably. Typical figures for the height of the F_2 layer are 250–350 km in the winter and 300–500 km in the summer.

The behaviour of the ionosphere when a radio wave is propagated through it depends very much upon the frequency of the wave. At low frequencies the ionosphere acts as though it were a medium of high electrical conductivity and reflects, with little loss, any signals incident on its lower edge. It is possible for a VLF or LF signal to propagate for considerable distances by means of reflections from both the lower edge of the ionosphere and the earth. This is shown by fig. 22.2. The wave suffers little attenuation on each reflection and so the received field strength is inversely proportional to the distance travelled.

For radio signals in the MF band the D layer acts as a very lossy medium; MF signals suffer so much loss in the D layer that little energy reaches the E or F layers. At night-time, however, the D layer has disappeared and an MF signal will be refracted by the E layer and perhaps also by the F layer(s) and returned to earth.

With further increase in frequency to the HF band, the ionospheric attenuation falls and the E and F layers provide refraction of the sky wave. At these frequencies the D layer has little, if any, refractive effect but it does introduce some losses.

The amount of refraction of a radio wave that an ionospheric layer is able to provide is a function of the frequency of the wave, and at VHF and above useful refraction is not usually obtained. This means that a VHF or SHF signal will normally pass straight through the ionosphere.

2 Fading

Fading, or changes in the amplitude of a received signal, is of two main types: general fading, in which the whole signal fades to the same extent; and selective fading, in which some of the frequency components of a signal fade while at the same time others increase in amplitude.

As it travels through the ionosphere, a radio wave is attenuated, but since the ionosphere is in a continual state of flux the attenuation is not constant, and the amplitude of the received signal varies. Under certain conditions a complete fade-out of signals may occur. General fading can usually be combated by automatic gain control (a.g.c.) in the radio receiver.

The radio waves arriving at the receiving end of a sky-wave radio link may have travelled over two or more different paths through the ionosphere (fig. 22.3a). The total field strength at the receiving aerial is the phasor sum of the field strengths produced by each wave. Since the ionosphere is subject to continual fluctuations in its ionization density, the difference between the lengths of paths 1 and 2 will fluctuate and this will alter the total field strength at the receiver. Suppose, for example, that path 2 is initially one wavelength longer than path 1; the field strengths produced by the two waves are then in phase and the total field strength is equal to the algebraic sum of the individual field strengths. If now a fluctuation occurs in the ionosphere causing the difference between the lengths of paths 1 and 2 to be reduced to a half-wavelength, the individual field strengths become in antiphase and the total field strength is given by their algebraic difference.

The phase difference between the field strengths set up by the two waves is a function of frequency and hence the phasor sum of the two field strengths is different for each component frequency in the signal. This means that some frequencies may fade at the same instant as others are augmented; this can result in distortion of received signals.

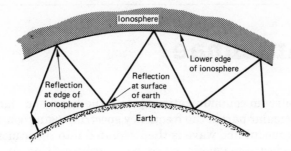

Fig. 22.2 Multi-hop transmission of a low-frequency wave

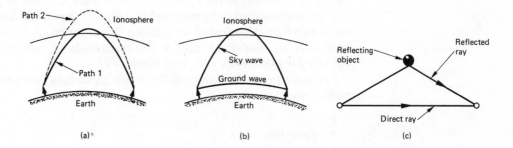

Fig. 22.3 Multi-path propagation

Selective fading cannot be overcome by the use of a.g.c. in the receiver but several methods of reducing it do exist. For example, the use of frequencies as near to the maximum usable frequency as possible, the use of a transmitting aerial that radiates only one possible mode of propagation, the use of single-sideband or frequency-modulated systems, or the use of such specialized equipment as Lincompex (linked compression and expansion). Selective fading of the sky wave is most likely when the route length necessitates the use of two or more hops. Suppose, for example, that a two-hop link has been engineered. Then, because of the directional characteristics of the transmitting aerial, there may well also be a three-hop path over which the transmitted energy is able to reach the receiving aerial.

Selective fading can also arise with systems using both surface and space waves. In the daytime the D layer of the ionosphere completely absorbs any energy radiated skywards by a medium-wave broadcast aerial. At night the D layer disappears and skywards radiation is returned to earth and will interfere with the ground wave, as shown in fig. 22.3*b*. In the regions where the ground and sky waves are present at night, rapid fading, caused by fluctuations in the length of the sky path, occurs.

Fig. 22.3*c* illustrates how multi-path reception of a VHF signal can occur. Energy arrives at the receiver by a direct path and by reflection from a large object such as a hill or gasholder. If the reflecting object is not stationary, the phase difference between the two signals will change rapidly and rapid fading will occur.

23 Antennae

In a radio communication system the baseband signal is positioned in a particular part of the frequency spectrum using some form of modulation. The modulated wave is then radiated into the atmosphere in the form of an electro-magnetic wave by a transmitting antenna (or aerial). A transmitting antenna may handle many kilowatts of power and has to be carefully matched to its feeder cable to ensure maximum power input.

For a radio signal to be received at a distant point, the electro-magnetic wave must be intercepted by a receiving antenna. A receiving antenna may only be concerned with a few milliwatts of power; its priority is usually for maximum gain and directivity.

For most practical purposes there are nowadays four common types of radio antenna:

1) Yagi
2) Rhombic
3) Log-periodic
4) Parabolic

Brief descriptions are given below of these different types of antenna. All may be used either for transmitting or for receiving; parabolic dishes are in fact often used both for transmitting and receiving at the same time, using different frequencies.

1) *The Yagi* This antenna, named after its Japanese inventor, is basically a conductor whose electrical length is one-half the wavelength at the desired frequency of operation, and is centre-fed. This is the basic $\lambda/2$ dipole, shown in fig. 23.1.

The radiation patterns, or the graphical representation of the way in which electric field strength produced by an antenna varies at a fixed distance from the antenna, in all directions in the plane concerned, are given in fig. 23.2 for the horizontal plane (a circle) and for the vertical plane.

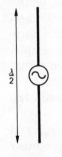

Fig. 23.1 The $\lambda/2$ dipole

Fig. 23.2 Radiation patterns of a vertical $\lambda/2$ dipole: (a) horizontal plane pattern; (b) vertical plane pattern

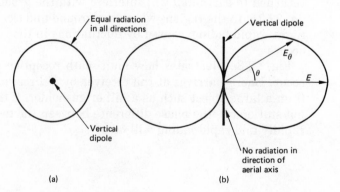

(a)

(b)

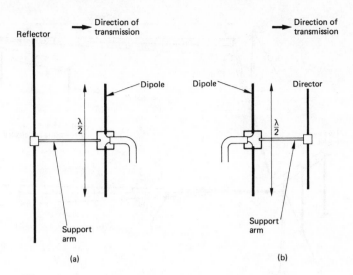

Fig. 23.3 λ/2 dipole with (a) a reflector and (b) a director

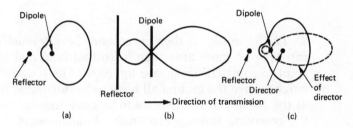

Fig. 23.4 Radiation patterns: (a) λ/2 dipole and reflector, in equatorial plane; (b) λ/2 dipole and reflector, in meridian plane; (c) λ/2 dipole, reflector and director, in equatorial plane

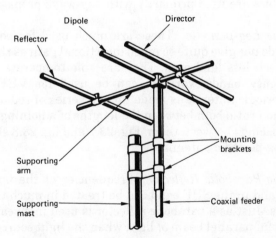

Fig. 23.5 A practical Yagi aerial

Increases in directivity of a λ/2 dipole may be obtained by adding parasitic elements called reflectors (about 5% longer than the dipole) and directors (about 5% shorter), mounted at carefully calculated distances from the dipole itself (figs 23.3 and 23.4). More directors, all mounted on the side of the dipole facing the direction of transmission, will give more directivity. Yagi antennae are widely used for the reception of TV broadcast signals (fig. 23.5).

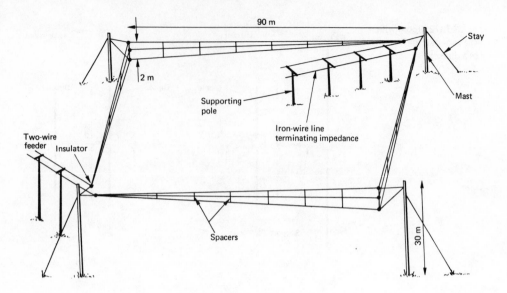

Fig. 23.6 A practical rhombic aerial

2) *The Rhombic*   A rhombic antenna is very much larger than a Yagi. It can take up more area than a football field and is used principally for point-to-point HF links (see fig. 23.6). The lengths of each arm and the height above the ground all have to be calculated for optimum efficiency at the particular frequencies to be used and the distance and bearing of the receiving station. The angle of elevation of the main beam (determined largely by the height of the antenna) is important because rhombics are used primarily with sky-wave propagation systems.

3) *The Log-periodic*   These are much used for point-to-point services. They do not give quite so much directional gain as rhombics but they take up much less land area. They are able to operate efficiently over wide frequency bands; variants can be used for VHF or HF services. A log-periodic antenna is made up of a series of rod or wire dipoles with a common ratio both between the lengths of adjoining rods or wires and for the spacings between them. Fig. 23.7 and fig. 23.8 show typical VHF and HF log-periodic antennae.

4) *The Parabolic Reflector*   Frequencies at the upper end of the UHF band and in the SHF band can be treated in much the same way as light beams. Just as a parabolic reflector is used in a searchlight to produce a powerful parallel beam of light when the light source is located exactly at the parabolic focus, so a parabolic reflector (often called a dish) can be used to provide a very directional high-gain antenna with the radio energy concentrated into a parallel beam. Dish diameters vary from about 20 cm for an internal antenna for a domestic TV receiver up to more than 30 metres for ground stations working to geostationary satellites (fig. 23.9).

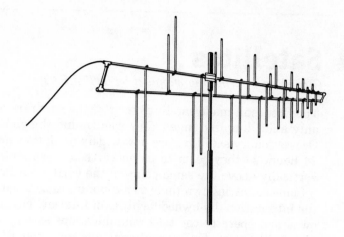

Fig. 23.7 A practical log-periodic aerial

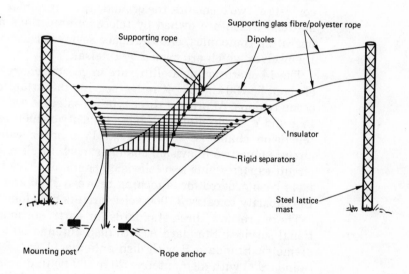

Fig. 23.8 A high-frequency log-periodic aerial

Supporting rope

Supporting glass fibre/polyester rope

Dipoles

Insulator

Rigid separators

Steel lattice mast

Mounting post

Rope anchor

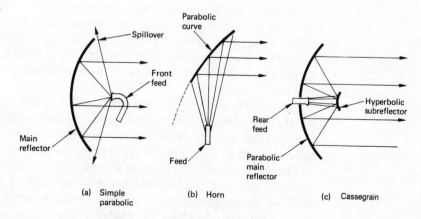

Fig. 23.9 Three versions of the parabolic antenna, showing placement of RF feed and the path of reflected energy. The Cassegrain is the most common for satellite communications

Spillover

Parabolic curve

Front feed

Main reflector

Feed

Rear feed

Parabolic main reflector

Hyperbolic subreflector

(a) Simple parabolic

(b) Horn

(c) Cassegrain

24 Satellites

(Based on information supplied by INTELSAT.)

The first man-made satellites to orbit the earth travelled at altitudes of only a few hundred miles; they went round the globe in about an hour. Geostationary satellites are now widely used; they go round the globe in 24 hours so they seem to us on earth to keep station some 35 780 km vertically above the same point on the earth's surface.

Some countries own their own communications satellites but most use the internationally-owned facilities of Intelsat. The Intelsat organization owns and operates the telecommunications satellites used by countries all over the world for international, and some domestic, communications. At present, 104 countries are members of Intelsat. Together they operate 253 communications antennas at 208 earth station sites. The system consists of two elements: the ground segment composed of the many earth stations which are owned by telecommunications administrations and operating companies, and the space segment consisting of the satellites themselves which are owned by Intelsat.

The 14 operational satellites are in geostationary orbits at an altitude of about 35 780 km (22 240 miles) serving the Atlantic, Indian and Pacific ocean regions. At present three types of satellite are used: the Intelsat IV, IVA and V. Each Intelsat IV has a capacity of 4000 voice circuits plus two television channels; each Intelsat IVA has a capacity of 6000 voice circuits plus two television channels; each Intelsat V has a 12 000 voice circuit capacity plus two television channels. Five Intelsat VI satellites have been ordered for launching between 1986 and 1991, and will have the capacity to carry 33 000 voice circuits plus four television channels.

There are now three standards for earth stations operating international services: Standard A with dish antenna 30 m (100 ft) or more in diameter, Standard B with dish antenna 11 m (36 ft) in diameter, and Standard C with dish antenna 20 m (60 ft) in diameter.

The Intelsat IVA satellites are large spin-stabilized cylinders, 2.81 m in diameter and weighing 825 kg, with solar cells all around the outside of the cylinder to provide power. The antennae are mounted at one end on a "de-spun" shelf to ensure that they always point toward the earth. The Intelsat V satellites weigh approximately 1011 kg and have their solar cells on extended arms, giving a "wing-span" of 15.7 m.

The Intelsat VI satellites will weigh about 2000 kg, and will be about 12 m high and 4 m across. Their solar panels will generate 2200 watts of power.

Intelsat IV and IVA satellites use 6 GHz links up to the satellite and 4 GHz links down to earth. Their effective bandwidth is 80 MHz. Intelsat V satellites use the same 4 GHz and 6 GHz bands and also use 14 GHz up-links and 11 GHz down-links. Their effective bandwidth is 2300 MHz.

The three maps show Intelsat's three regions and the locations of earth stations working to the satellites serving each of these ocean regions; each of these satellites is in geostationary orbit parked in space at equatorial latitude above the centre of the region concerned.

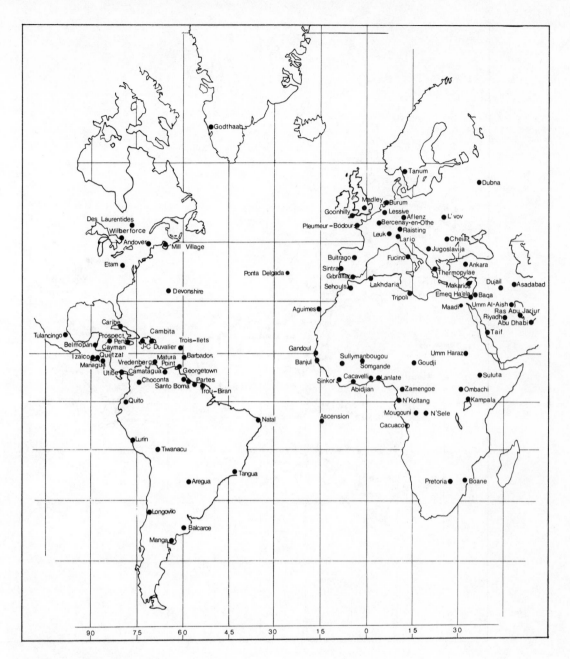

Fig. 24.1 The Intelsat
system: Atlantic Ocean
region

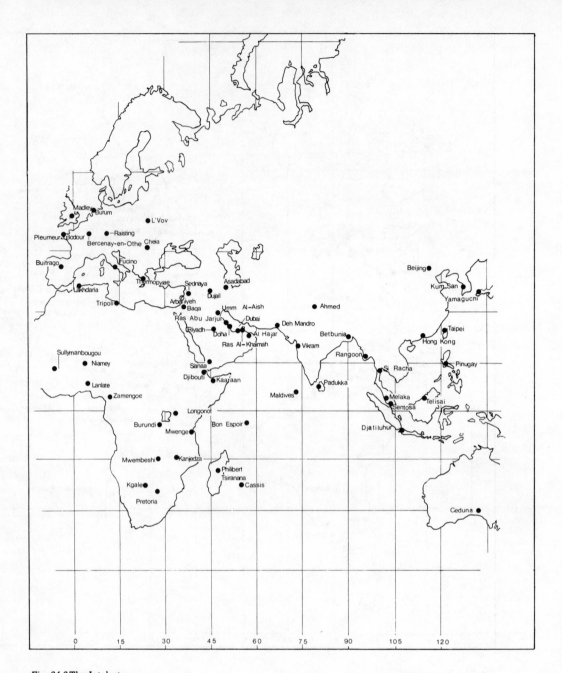

Fig. 24.2 The Intelsat
system: Indian Ocean
region

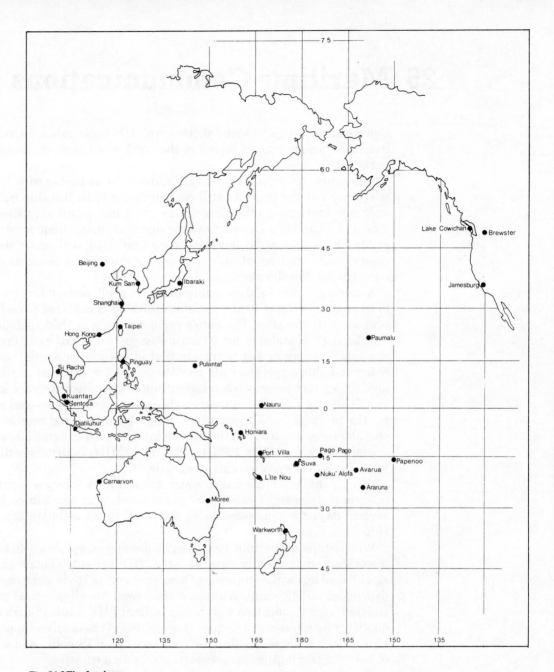

Fig. 24.3 The Intelsat system: Pacific Ocean region

25 Maritime Communications

Conventional long-distance "short-wave" HF radio relies on reflection from the various ionised layers in the earth's atmosphere, as described in Section 3.

Some ships are equipped with HF radio-telex and some with HF radio telephony but the majority still rely on morse code; if a ship out at sea wishes to transmit a telegraphic message, it has to call out using morse code and establish a two-way radio connection, using frequencies appropriate for the time of day and the distance involved, with one of the many coastal radio stations which have been established by telecoms administrations all over the world.

A message from land to a ship presents a bit more difficulty: it has to be routed to the coastal radio station which is expected to be in radio contact with the ship. The ship's radio call sign is then included in a "traffic list" of stations for which messages are held, awaiting transmission. This traffic list is transmitted at regular intervals, say every hour, and ship's operators all over the world are expected to listen out and, if they spot their own call sign in one of these lists, they are expected to call the coastal station concerned—again using morse—and arrange for the message to be transmitted. Ship's operators and coastal station operators also listen out all the time on internationally agreed frequencies (usually 500 kHz, 8 MHz, 12 MHz, 16 MHz, 22 MHz) for distress calls; these take absolute priority over all other calls.

Clearly this is a situation in which immediate delivery of a particular telegraphed message cannot be guaranteed, and sometimes it takes several days for a message to be delivered to its ultimate destination ship.

When ships get within line-of-sight distance (say about 40 km) of a major port, it is usually possible for a VHF radio telephone call to be established but while ships were "over the horizon" they were completely dependent on HF radio. Without it they were for all practical purposes isolated from the modern world. And ordinary HF radio, thanks to poor-quality transmission and frequency congestion, is basically a support function only, a reserve life-line for emergencies (SOS calls) and a means of providing for minimum necessary information transfer.

The dawning of the satellite age led, in 1979, to the formation of the International Maritime Satellite Organizations, Inmarsat. Operational services were opened in 1982 and Inmarsat now provides first-class telecommunication services, 24 hours a day, 7 days a week, to all suitable equipped ships in all the oceans of the world.

Fig. 25.1 shows the coverage of the three ocean zones; the Operations Control Centre (OCC) is in London, while the Network Coordination Stations for the three ocean areas are in Ibaraki, Japan (Pacific), Southbury, USA (Atlantic), and Yamaguchi, Japan (Indian Ocean).

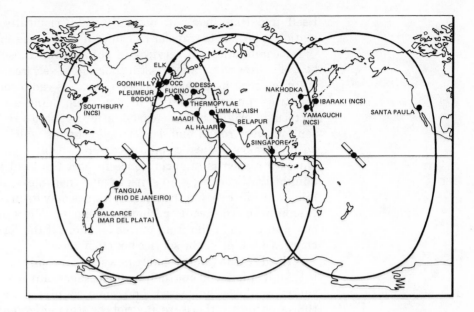

Fig. 25.1 Inmarsat's near-global coverage (OCC = operations control centre; NCS = network coordination station; coastal earth stations ●)

A ship's earth station is a comparatively small affair quite unlike its terrestrial cousins with their 30-metre-diameter dishes. The ship's antenna is mounted on a sophisticated stabilization system in order to keep the dish pointing straight to the particular satellite concerned however violently the ship rolls, and the whole assembly is covered by a lightweight waterproof dome. Typical dishes are only 0.9 metre in diameter. The whole dome unit is less than 2 metres high and 1.5 metres in diameter, and in weighing less than 200 kilograms can easily be mounted on a ship's mast out of everyone's way, even on quite a small ship. About 3000 ships are already equipped with stations of this type.

Setting up a telephone call to a passenger on a big ship, or to the captain or a crewman on a small ship, is nowadays almost the same as dialling one's own local neighbour. The CCITT (International Consultative Committee for Telephony & Telegraphy) has drawn up an internationally agreed numbering plan with digits to indicate the country of registration of the ship as well as the ship itself, because someone, somewhere on land, has to be willing to pay the bills for any calls made. Procedure is just as for a normal international self-dialled call.

Now that good-quality voice-grade circuits can be established by satellite to a ship, all the many telecommunications services and terminals which are already in use in terrestrial offices can now be made available for use in ships, including telex, teletex (supertelex), and facsimile services. One other service which is already available on land, the viewdata (Prestel, Telidon, etc.) service (see Section 48) is not yet widely available on ships but is the subject of a great deal of development work. The availability of pages of easily readable weather forecasts could, for example, be extremely useful on board ships as soon as an economic way can be designed either to store the pages of weather information on the ship itself, updating them automatically and periodically, or for a land-based station to be able to transfer this information to the ship on request.

Telecommunication services in ships—even those equipped with the best of ships earth stations—still suffer, however, from administrative difficulty: as soon as a ship reaches a port, its radio services are, in many countries, physically sealed to stop them being used. This is probably historic in origin, because early designs of ship's HF radio transmitters sent out very rough high-power radio signals which caused interference with ordinary domestic radio receivers within a mile or so of the port. Another reason is financial: once in port the local telecommunications administration expects all ships to send their messages through the local telegraph office so that the telegraph office may itself collect some revenue. A ship in harbour could perhaps arrange to collect outgoing traffic from other ships and even from offices on shore if the local telegraph office's charges are high or the service not efficient.

A third reason (and probably nowadays the only one of real importance) is that, if ships send out their own messages, any local censorship organization would be by-passed, and some countries are very touchy about this, especially if there are any riots or other emergencies in the area.

But the arrival of the satellite age has at last enabled offices in ships to be provided with all the advanced services which are already available for offices on dry land.

26 Mobile Radio Systems

A useful extension to any national telephone system is a Mobile Automatic Telephone System or MATS. The aim should be for each mobile station—a radiophone in a car—to have its own individual number so that it can for all practical purposes be treated in the same way as a normal line-fed telephone. It must be able to dial calls not only to other mobiles in the same city but to ordinary lines (or mobiles) anywhere in the world, using normal long-distance and international circuits—and of course to be billed for all the calls initiated.

Until very recently the available technology has only made it possible for relatively few people to be permitted to use car phones of this type. The reason for this is that radio coverage has been provided by few relatively high-power radio transmitters and associated very sensitive receivers. Each call to a mobile uses two frequencies (one in, one out, to give both-way conversation), so a call between two mobiles would use four frequencies, each with a bandwidth of about 25 kHz. Thus, a total bandwidth of 100 kHz would be "frozen" and unavailable for other users over the whole service area of the radio stations providing the coverage— and the bandwidth available for use by mobile services is severely limited. This is why there were reputed to be almost as many potential

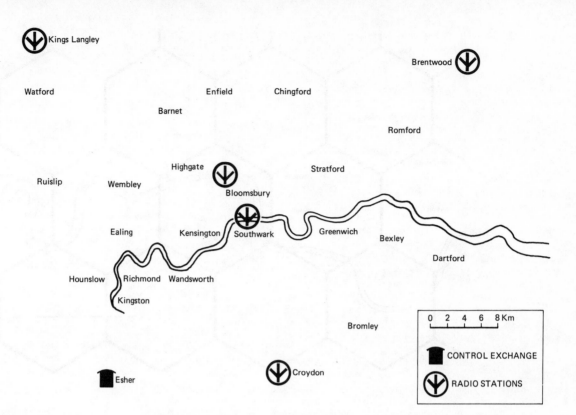

CONTROL EXCHANGE

RADIO STATIONS

Fig. 26.1 London's mobile radio service: the pre-1985 high-power stations

users on the waiting list for radiophone service in London as there were actual users—and why the actual users sometimes got a comparatively poor grade of service.

The technological developments which have made improvements now possible are in part dependent on the invention of the microprocessor and in part a spin-off from military radio design.

During the Second World War, to change a radio's operating frequency was a tedious business involving manual re-tuning of circuits and readjustment of antennae. Both sides monitored the other's radio transmissions and did their best to work out which formations were in which locations by studying operating practices and traffic patterns. Strict radio silence was of course enforced during the build-up period before major operations, and in an attempt to confuse the other side all the radio links in a formation usually changed frequency at irregular intervals, about once a week—more frequently than this was administratively impossible with the techniques then in use. Current-generation military radios are very much more difficult to monitor; they can change frequencies many times a second under the control of microprocessors and pseudo random number generators; these frequency changes are made simultaneously at all the stations on each link. This same technique is now available for civilian use in a much simplified form.

Instead of covering the whole area by high-power fixed radio stations (fig. 26.1) the area is divided up into small cells, only a few kilometres across (fig. 26.2). As the subscriber density for mobile telephones increases, the cell pattern can be changed to accommodate it. First of all,

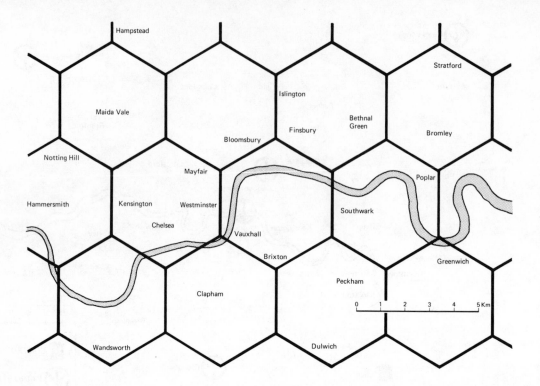

Within the figure, the following cell labels appear:

Hampstead

Stratford

Islington

Maida Vale

Bethnal
Green

Finsbury

Bromley

Bloomsbury

Notting Hill

Mayfair

Poplar

Hammersmith

Kensington

Westminster

Southwark

Chelsea

Vauxhall

Brixton

Greenwich

Peckham

Clapham

0 1 2 3 4 5 Km

Wandsworth

Dulwich

Fig. 26.2 Cellular radio: a possible scheme for Central London. Each cell has its own low-power transmitters and sensitive receivers

more working channels are brought into service in the cell, then the size of the cell is decreased, either by the creation of new cells or by changing the boundaries with adjoining cells. A single multi-channel transmitter and receiver is located in each cell. The transmitters use such low power that the same radio frequency bands can be used again and again, in separated cells. permitting the number of simultaneous users to jump up from only hundreds to many thousands.

As a mobile station—your car with its radiophone—moves across the city, the signal-to-noise ratio of its received signal varies at the cell station to which it is working. If this ratio becomes worse than a pre-determined level, the microprocessor in the cell station advises the central processor which manages the whole system. This checks with the stations in each of the adjoining cells to find which one of them is now receiving the signal well from your mobile. Control of the call is then transferred to the station in this other cell, the radio set in your car changes its transmit and receiving frequencies to interwork with its new master, and the telephone exchanges directly associated with each cell station switch the circuit on which you were talking from your original cell to your new cell station. All this happens in thousandths of a second, without you being aware of any changes at all.

There are several competitors' schemes now on the market, all of which enable cell switching of this type to be provided. They differ mainly in the methods of control and of signalling between the different base stations and mobiles. There is no doubt that by 1990 there will be a great many more phones in cars than there were before cellular radio was invented.

Part E Maintenance and Operation

27 Centralized Maintenance Systems

Most modern telephone exchanges require no maintenance adjustments: there are in general no moving parts so there is nothing to lubricate, nothing to clean, polish or re-adjust. When networks change over from electro-mechanical space-division switching and analog transmission systems to computer-controlled solid-state time-division switching and digital transmission systems, all the administrative, operational and maintenance functions can be streamlined.

Operational and maintenance services normally comprise four basic functions:

1) Service management, i.e. control of service offered to subscribers.
2) Maintenance management, i.e. the maintenance and repair of exchanges, external plant, transmission equipment, subscribers' apparatus. Software may also need attention.
3) Network management, i.e. management and supervision of traffic flow; short-term and long-term planning.
4) Call accounting, i.e. control of all aspects of recording information in relation to the bills to be sent to subscribers.

Because there is nothing to adjust and the number of faults is likely to be very small, it is not usually necessary for modern exchanges to be manned by permanently stationed maintenance staff. Most national administrations now therefore centralize their maintenance activities, merely sending technicians out to visit exchanges to change faulty items. Most types of modern exchange incorporate diagnostic programmes which not only localize their own faults (and give a print-out describing the card or unit which is faulty and should be changed) but rearrange the traffic handling pattern of the exchange so that any such faulty section is by-passed. There is usually sufficient redundancy built into the exchange design to maintain traffic handling capacity during all except the most severe faults; it is indeed probable that most subscribers on an exchange will not even know that a fault has developed, been localized and cleared unless they happen to see the technician's van parked outside the exchange building.

This centralization of maintenance efforts means that in the unlikely event of an exchange developing particularly severe problems, it is possible to bring greater resources to that exchange to ensure that normal service is resumed with the minimum of delay. The organization of the maintenance effort into a single large group also allows expertise to be shared and ensures that individuals are continuously exercising and improving these skills.

Most administrations complement their centralized maintenance operations by centralizing all repair activities also. It is rare for maintenance staff at a modern exchange to be required to test and repair faulty printed circuit boards. First line maintenance normally takes the form of substitution of a new printed circuit board using information provided by the exchange itself and sent as a data message to the remote operations and maintenance centre. The replaced unit will then normally be sent to a specially equipped repair centre for repair. Here it must be borne in mind that many printed circuit boards now incorporate microprocessors so if a board is suspected of being faulty it has to be tested by a sophisticated and expensive computer-controlled test set. This test set will have been programmed to test all the functions of that particular board and of any other boards which might be sent to the repair centre for test and repair.

By the time any exchange has been brought into service, all its software should have been exhaustively tested, "de-bugged", and proved satisfactory. A master copy of each exchange software is usually held available for reference when necessary, in secure and non-volatile stores such as magnetic tapes, discs, drums or bubbles. It is common practice for wanted information to be called forward from one store to another, for example into an immediate access store. If software is suspected of being faulty (one plane of action may give a different answer from others), then the software concerned is usually replaced by a new version of the same software (in case the original has become corrupted), by calling forward from a "clean copy" held in a secure store. This procedure is called software reload or rollback. It is usual for there to be several levels of this, depending on the nature of the difficulty encountered—it would clearly be unnecessary for the whole of the exchange's software to be changed if changing only a single small module would eliminate the trouble. These software reload actions are normally carried out completely automatically at exchanges themselves, not on a centralized basis. Only if faults occur which cannot be cleared by changing over to a locally-available clean copy of the relevant software is reference made to a centralized maintenance organization. In real emergencies it is however often possible to connect an exchange not merely to a regional maintenance centre but to a national research laboratory or to the manufacturer's design engineers.

28 Telephone Tariffs and System Viability

1 Viability Study and Tariff Structures

Capital costs (line plant, switching equipment, buildings and land, etc.), financing costs (interest on loans, etc.) and operating costs (staff salaries, maintenance and power, etc.) all have to be taken into account when calculating the total expenditure involved in providing telephone services. If the administration is to stand on its own feet financially, all these costs must be recovered; it is normally also necessary for a profit to be earned.

In some countries a financial target is set by government; it is therefore necessary for a tariff structure to be established so that the telephone system will earn specified profits. In other countries the tariff structure is fixed on purely political grounds by deciding how much the people can reasonably be called upon to pay for telephone service. Some countries may in fact subsidize telephone services by providing them for citizens at less than their true total cost, as a social measure.

A viability study is usually prepared in draft form at an initial planning stage and then refined when major tenders have come in and been evaluated so that actual expenditure during a main development scheme can be considered, instead of basing calculations purely on estimates and forecasts.

Various techniques can be employed in measuring viability; discounted cash flow and return on capital techniques are the principal measures. A study will usually end by recommending rates of return considered likely to be economically acceptable.

Comprehensive telephone tariffs have to be designed to generate sufficient revenue to enable the system to meet its required rate of financial return. The objective is usually one of maximizing the use of the system while meeting revenue targets.

The advantages and disadvantages of the various methods of charging for local telephone service have to be considered; these methods include:

a) Full flat rate—no separate charges for calls made.

b) Limited flat rate—some calls charged for (e.g. a free call allowance or all calls within a specified area free of charge).

c) Message rate, untimed—charge per call irrespective of duration.

d) Timed measured rate—charge per unit of call duration. This is sometimes called usage-sensitive charging.

Combinations of these different methods are possible, particularly in a large system.

The most suitable method for implementation then has to be selected and a network call charging plan produced which will conform with the technical capabilities of the system. Tariff proposals not only have to cover the charges for exchange line service and for calls, but must also include the charges for additional apparatus such as private branch exchanges and extension telephones. These charges normally have to be cost related so that no one service is "carried" by others.

2 *Metering and Billing*

Administrations which charge for local calls usually meter them by associating a simple cyclometer type meter with each exchange line. Meters are normally mounted about 1000 to a rack; they are read (sometimes photographically) and bills prepared on the basis of current reading less previous reading (less the number of test pulses used in the period). Newer exchanges can provide the same meter readings by storage of totals on a software basis; the accumulative totals can be read out whenever necessary.

There are three basic methods of using ordinary subscribers' meters for trunk and junction call charging:

1) Multimetering: when the called subscriber answers, the calling subscriber's meter is stepped up a specific number of times, depending on the exchange to which the call is routed. No timing is involved.

2) Repeat Multimetering: when the called subscriber answers and every 3 minutes thereafter, meters are stepped up by a fairly rapid train of pulses, e.g. six pulses (in 3 secs) (if the charge is 6 × local call fee). Pulses are generated within the exchange and operate straight to the meters. The particular pulse rate given depends on the exchange to which the call is routed.

3) Periodic Pulse Metering: PPM involves the selection of a pulse rate which is dependent on distance, e.g. calls from London to France could be pulsed once every 7.2 seconds, once every 4.8 seconds to the USA, once every 2.4 seconds to Nigeria, once every 2 minutes for a daytime local call, and so on. These timing intervals can rapidly be changed, making fine control of revenue possible.

All three of these systems involve bulk billing. The same meter is used for local calls as for trunk calls so there can be no separation of charges.

The ordinary electro-mechanical subscriber's meter itself will operate satisfactorily about 3 times per second, and metering in software could be many times faster than this, but fee determination is often carried out at an exchange other than that of the originating subscribers, so pulses have to be sent over junctions back to the originating exchange. Metering over junction equipment cannot be expected to operate without distortion more rapidly than 1 pulse per second.

Bulk billing has advantages of simplicity for the administration but there are disadvantages also, both for the customer and the administration:

a) International Subscriber Trunk Dialling (ISTD) and national STD charges have to be closely related to local unit-call charges, so rates cannot be changed (e.g. to take account of currency exchange rate changes) unless pulsing equipment has suitable spare outputs readily available or can readily be changed or modified.

b) Limitation to a maximum of 1 pulse per second (i.e. a maximum charge of 60 local call fees per minute) is insufficient to cover costs in areas where the local call tariff is held down by Government action and where distances are large and there are only small numbers of circuits in traffic groups. In Europe and in the USA, circuit groups are big enough to give real economies of scale, whereas in some territories demand is relatively small and can only justify the use of comparatively expensive 6 or 12 channel systems.

c) The presentation of an itemized trunk call account is expected by customers who are used to North American practice.

Bulk billing can produce a great many complaints by customers (and resultant non-revenue earning investigations by administrations), especially when expensive international calls can be dialled direct.

Several methods of automatically preparing itemized bills for self-dialled trunk calls have been developed. These can be divided broadly into systems in which some sorting is done at the origin and those in which sorting is done at a computer centre which can be remote from the exchange. There is some confusion about definitions but, in general, Automatic Message Accounting or AMA refers to the whole procedure, toll ticketing to a system with local sorting, and Centralized AMA or CAMA to a system with remote sorting:

a) Toll Ticketing: all the information necessary to prepare the bill for each call is punched on a single card or recorded in one stretch of magnetic tape (e.g. calling number, called number, time on, time off, etc.).

b) Centralized Automatic Message Accounting (CAMA): all required information about calls is recorded as it happens (there is usually a small temporary data store only), usually on magnetic tape which then has to be processed by a computer to link together "time on" for a particular call with "time off" for the same call so that chargeable time may be calculated.

29 Forecasting Future Demands

It was undoubtedly a telephone system planner who first said that forecasting was always difficult, especially forecasting the future. Telephone people are especially sensitive to forecasting difficulties because someone always has to take the blame when telephone service cannot be provided in any given location immediately it is requested. Unfortunately even what seems to be a simple request for service sometimes behaves like the last straw and breaks the camel's back.

In most countries it takes between five and seven years to get a new telephone exchange into service, starting from the selection and purchase of the site (sometimes this means that a lengthy public enquiry has to be held), the design and construction of the building, the design, manufacture and installation of the switching equipment itself, the selection and training of all staff needed, the construction of all necessary manholes and duct routes so that cables can be fed into the new exchange, the planning, purchasing, laying and jointing of all these new cables. And many existing exchanges will be affected by any proposed new exchange and will themselves need extra equipment to be designed, purchased, installed and tested. So the time passes very rapidly.

Planning a new telephone system normally begins with the preparation of a set of large-scale maps of the area concerned showing every plot of land and every building. The area is divided on these maps into blocks about 100 metres by 100 metres in size. All available information is collected together about possible new road construction schemes, housing schemes, office blocks, long-term municipal development plans, and so on. Visits are then paid by skilled staff to each of the blocks. Starting with up-to-date information about the number of existing telephone lines, telex lines, etc., in each of the buildings in the block, plus the numbers on waiting lists for service, estimates are made of the probable numbers of lines likely to be needed in 5, 10, 15 and 20 years' time, based on a study of all available planning information. These forecast figures are duly entered in boxes on the master map.

Study of these figures will indicate the preferred location for an exchange on the basis of it being the "copper centre": the location which could serve all the forecast lines at minimum cable cost. The practical copper centres often move across a city with the passage of time, e.g. with the development of an industrial estate in the suburbs. Modern transmission and switching practices, such as the use of digital exchanges and PCM systems to serve remote concentrators located in outlying parts of a city, nowadays make the theoretical copper centre of less direct relevance when a new site has to be selected but, if there is a choice of sites, the different total costs of cabling needed for each possible exchange site all have to be considered before reaching a decision. Civil works, including the laying of duct routes, the construction of manholes, etc., are usually planned on the basis of 20 years' life; it becomes extremely expensive if roads have to be excavated and surfaces reinstated every few years. Primary distribution cables are then pulled in to these newly laid ducts; usually enough cable pairs are provided initially to meet the 5 year forecast of demand for the area concerned. Secondary distribution cables, the smaller cables feeding from flexibility cabinets out to distribution points, are often directly buried in the ground, i.e. without using ducts. In such cases an attempt is usually made to provide sufficient cable pairs to meet the demand forecast for at least 15 years ahead. If the information available to the original planners was incomplete, or if major changes of land use are agreed and locally implemented at short notice, it is no wonder that there are sometimes insufficient cable pairs available to meet current demands for service.

The expansion of complete national systems can best be looked at on an overall basis. If demand for service is plotted against time using ordinary arithmetic scales, it is usual for this curve to take one of a number of shapes, depending on many economic factors and on the length of the period being studied (see fig. 29.1):

Fig. 29.1 Types of demand curve

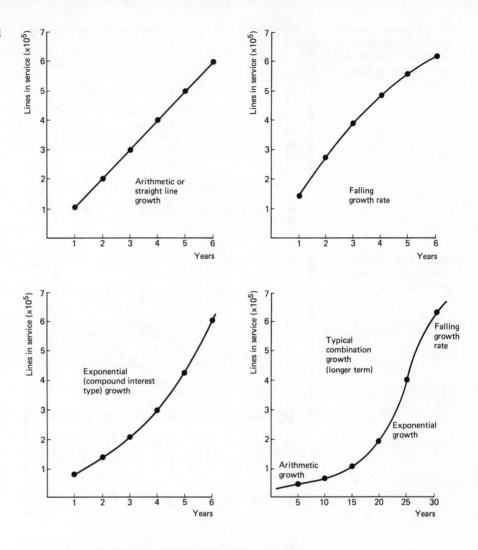

a) Straight-line growth rate.
b) Exponential or compound interest type growth rate.
c) Falling-off in growth rate.
d) A combination of all three of these.

Most developing countries experience an exponential or compound-interest type of growth of demand; it will be seen that it is extremely difficult to use a graph of this nature to extrapolate forward to give forecast figures for later years, since the line becomes almost vertical (fig. 29.2). Here the usefulness of semi-logarithmic graph paper becomes apparent: when compound-interest type growth is plotted on this paper a straight line results which can legitimately be extended forward to provide forecast figures for future demand (fig. 29.3).

Planners often have to decide which way to tackle a job in order to obtain the best value for money. One simple example is given in fig. 29.4—here the choice shown was between

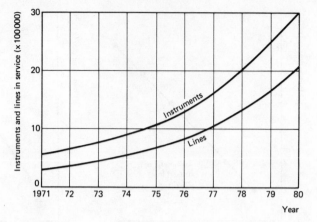

Fig. 29.2 Typical figures for the growth of an actual national telephone system: growth plotted on linear scales

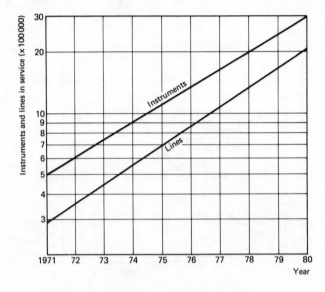

Fig. 29.3 As fig. 29.2 but growth plotted on semi-logarithmic paper

1) Installing new analog exchanges.
2) Installing new digital exchanges.
3) Installing new digital exchanges using remote concentrators.

The third alternative clearly provided all the necessary service features at a capital cost of only 79% of the original "traditional" analog scheme.

In an ideal world, good planning would be followed by the installation of just enough equipment to satisfy all demands for service, as they occur. In our real and imperfect world there are however usually a few hiccups. Actual development on the ground all too often requires different telecommunication services (or more of them) than was envisaged by the planners. Also it is not often possible to provide telecommunications equipment in small packets, keeping just ahead of demand. For example, if a trench is to be excavated along a road and ducts and manholes constructed it is usually economic—and certainly much preferred by residents—for such works to be done not more often than once in twenty years or so. This necessarily means that many of the ducts laid will be

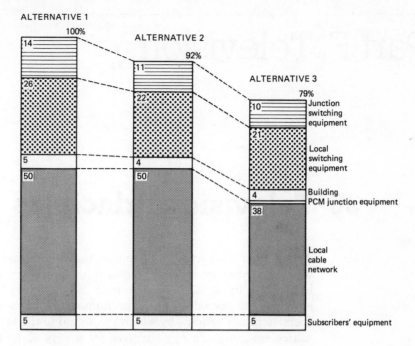

Fig. 29.4 Cost comparisons of various network planning alternatives

empty and not earning their keep for several years. The planner's aim must always be to keep this "burden of spare plant" as low as possible and to use technological improvements and advancements not only to improve services provided for customers but also to reduce the capital and running costs of the system.

Part F Television

30 Television Principles

Actors in a film appear to move smoothly across the cinema screen because a number of "still" pictures are presented on the screen to the human eye in rapid succession, each "still" picture being slightly different from the preceding one. The human eye has a characteristic called "persistence of vision", by which the signal to the brain caused by a light source reaching the eye survives for a very short time after the light source is removed. If "still" pictures are presented one after another to the human eye at a rate of more than 16 per second, an illusion of a moving scene is created but in some circumstances there can be significant flicker unless the rate is increased. A television system must therefore be designed to present pictures to the human eye from the TV receiver at a high enough rate to minimize or eliminate flicker.

A television system (fig. 30.1) uses one or more television cameras to convert the light energy of a natural moving visible scene, either in a television studio or outdoors, into electronic signals. Alternatively, the signals may be obtained from a video tape recorder, from telecine machines, or from slide scanners. These last two convert films or photographic slides into appropriate signals. These signals are usually conveyed by line to a television transmitting station where they modulate a carrier source, and the resultant vision-modulated carrier wave is passed to the transmitting aerial to be radiated in all directions as a broadcast vision signal.

At the same time, the sound energy information associated with the visible scene is picked up by a microphone and converted into an electronic signal which is also passed by line to the transmitting station to modulate a separate carrier source. The resultant sound-modulated carrier wave is then passed to the transmitting aerial to be radiated into the atmosphere along with the vision-modulated carrier wave.

Within a certain distance from the TV transmitting aerial, according to the amount of radio-frequency power radiated, a TV receiving aerial can pick up the combined vision- and sound-modulated wave to pass it to a TV receiver (fig. 30.2). The receiver amplifies the received signal, and then separates the vision and sound components after a demodulation process. The demodulated vision signal is passed to a cathode ray tube to reproduce as closely as possible the original visible moving scene at the

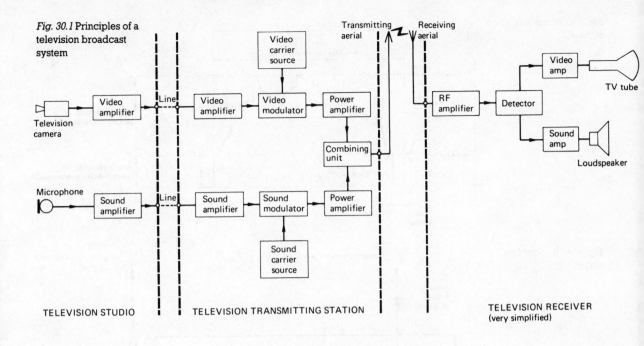

Fig. 30.1 Principles of a television broadcast system

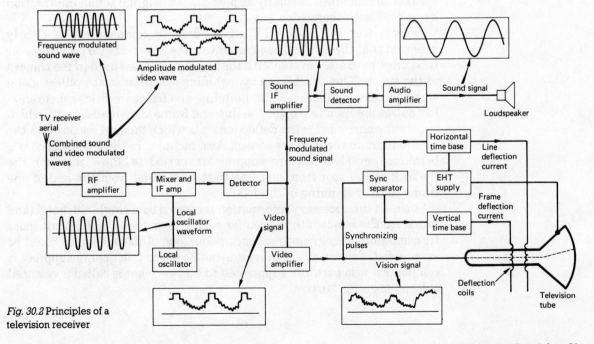

Fig. 30.2 Principles of a television receiver

Fig. 30.3 Principle of video signal

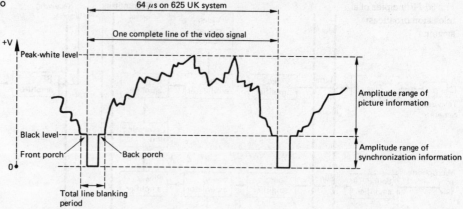

Fig. 30.4 The complete TV signal

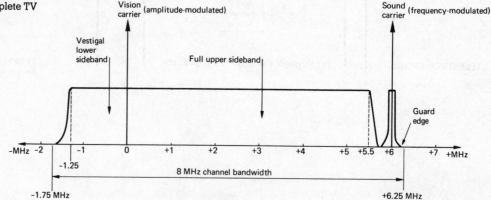

transmitting end. The demodulated sound signal is passed to a loud-speaker to reproduce as closely as possible the original sound information associated with the visible scene.

Several features of the TV signal deserve comment. It is clearly essential that the electron beam scanning the screen of every TV receiver tube must be exactly in step with the scanning process used in the camera at the studio. This is achieved by including synchronization pulses along with the video signal (fig. 30.3). Both line and frame (or field) synchronizing pulses are used to trigger the line and frame time-base circuits which supply the currents for the deflection coils which themselves position the spot of light on the receiver screen. Also included in the video signal is a blanking period or picture suppression period to allow time for the fly-back of the spot from one line to the next and from the end of one frame to the beginning of the next.

Since all the necessary information is carried by a single sideband (and to reduce the bandwidth needed for each transmission and permit more transmissions in a given frequency band), one of the sidebands could be suppressed. For various reasons it is not however practical to suppress it completely; it is partially suppressed to become what is called a vestigial sideband (see fig. 30.4).

31 Colour Television

1 Compatibility

When a colour television service is introduced in any country, it is clearly uneconomical to consider providing separate colour and monochrome services. Some viewers will obviously prefer to retain their monochrome receivers, if only because of cost. So it is necessary to provide a service that allows colour or monochrome viewing by choice of domestic receiver. This requires a signal from the television camera that can be received by either a colour or monochrome receiver, as required. Such a signal is called a compatible signal, and is produced by the television camera in two distinct parts:

> the *luminance* part
> the *chrominance* part

The luminance part contains the brightness information, similar to that already described for a monochrome-only system. The chrominance part contains the additional information needed for a colour system.

So, monochrome receivers use only the luminance part of the compatible signal, but colour receivers use both the luminance and the chrominance parts of the compatible signal.

There are currently three ways in which a compatible signal can be produced from a TV camera to provide the luminance and chrominance signals:

1) The NTSC system (National Television Systems Committee), developed and introduced in the United States of America in the early 1950s, and later also adopted in Canada, Japan and Mexico.

2) The PAL system (Phase Alternate Line), developed in West Germany from the NTSC system, and later also adopted in the United Kingdom and other European countries.

3) The SECAM system (Sequential Couleur a Memoire), developed in France, and later also adopted in East Germany, USSR and other countries in Europe and North Africa.

In all three systems, the luminance and chrominance information signals from the TV camera (fig. 31.1) are combined to form a compatible video signal (including line and frame synchronizing pulses), which then modulates a vision carrier frequency for transmission on a particular TV channel. The resulting modulated wave has to be accommodated in the same radio-frequency bandwidth allocated for a monochrome television channel.

The chrominance information signal is combined or interleaved with the luminance information signal at the transmitter by a process called encoding, and at the receiver the reverse process called decoding is used.

2 Colour

Every colour has three important properties: luminance (brightness), hue (colour), and saturation (amount of colour). The brightness or luminance information in a monochrome system is produced by a single electron tube in the TV camera. In a colour TV system, the camera contains at least three tubes, one each for the colours red, green and blue. These are the predominant or primary colours of the rainbow. By a reverse process, the colours of the rainbow can be recombined to form white light.

When red and green light are mixed, yellow is seen by the human eye. When green and blue are mixed, cyan is seen. When red and blue are mixed, magenta is seen. Yellow, cyan and magenta are called complementary colours.

By combining red, green and blue light in various intensities, it is possible to produce for the human eye any of a wide range of natural colours. The three primary colours of red, green and blue are extracted from the natural scene to be televised by using suitable light filters in front of each of the three camera tubes. The luminance and chrominance signals are then produced from the three primary colour signals in the encoding circuits, and combined to form the compatible video signal which is then passed to the modulator of the TV transmitter (fig. 31.2).

At a colour receiver, after demodulation, the compatible signal is applied from the decoding circuits to a three-gun cathode ray tube to reproduce the colour picture. At a monochrome receiver, after demodulation, the luminance part of the compatible signal is applied to a single-gun CRT to reproduce a black and white picture as already described.

3 A Colour TV System

The basic arrangement of a PAL colour TV system is illustrated in fig. 31.1 and is described here.

The luminance signal, which contains the brightness information, is produced by combining in the encoder the red, green and blue signals, from the TV camera tubes and filters, in the particular proportions required by the human eye to observe white light. This proportion is 30% red, 59% green and 11% blue.

The luminance signal is generally designated by Y, and can be represented by the equation

$$Y = 0.3R + 0.59G + 0.11B$$

This means that ordinary black-and-white TV sets can be used to receive a colour TV signal and will produce a satisfactory black-and-white picture from it.

The chrominance signal, the additional information for the colour signal, is placed within the normal bandwidth of the channel produced by modulating the allocated channel vision carrier frequency by the luminance signal.

The chrominance signal modulates a sub-carrier frequency, and this sub-carrier is suppressed at the transmitter. The sub-carrier frequency is approximately 4.43 MHz (fig. 31.3), and the bandwidth produced by the

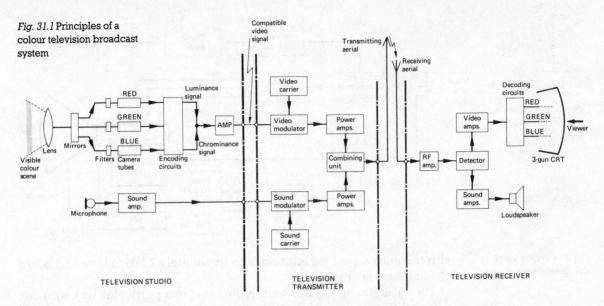

Fig. 31.1 Principles of a colour television broadcast system

TELEVISION STUDIO   TELEVISION TRANSMITTER   TELEVISION RECEIVER

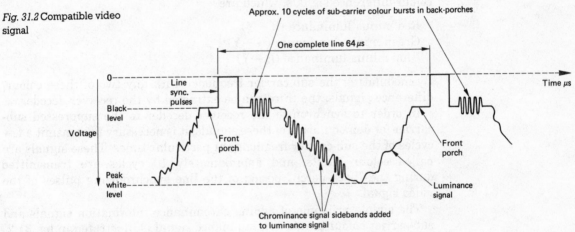

Fig. 31.2 Compatible video signal

Approx. 10 cycles of sub-carrier colour bursts in back-porches

One complete line 64 μs

Chrominance signal sidebands added to luminance signal

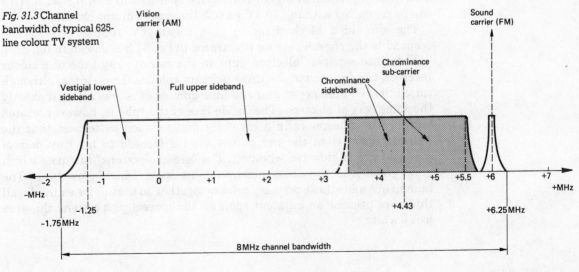

Fig. 31.3 Channel bandwidth of typical 625-line colour TV system

8 MHz channel bandwidth

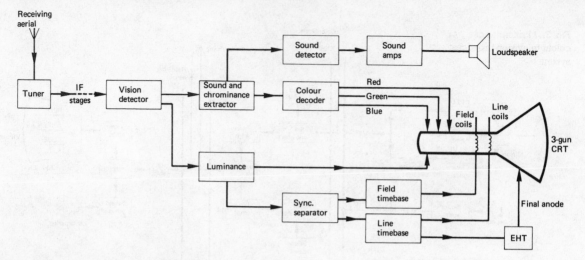

Fig. 31.4 Principles of
colour television receiver

chrominance signal modulation is approximately 1 MHz above and below
the sub-carrier frequency.

The chrominance signal itself is obtained in a particular way by using
colour-difference signals, which are

Red minus luminance (R − Y)
Green minus luminance (G − Y)
Blue minus luminance (B − Y)

By modulating the sub-carrier frequency with any two of these colour-
difference signals, the third can be extracted by the receiver decoder.

In order to synchronize the receiver decoder to the suppressed sub-
carrier for demodulation of these signals, it is necessary to transmit a few
cycles of the sub-carrier frequency at particular times. These signals are
called colour bursts, and approximately 10 cycles are transmitted
during the "back-porch" period of the line synchronizing pulses of the
video signal.

The simple principle of adding chrominance information signals and
sub-carrier colour bursts to a luminance signal is illustrated in fig. 31.2.
The resulting channel signal bandwidth is illustrated in fig. 31.3. This
can be compared with fig. 30.4 for a 625-line monochrome channel.

The simplified block diagram of a colour TV receiver which will
respond to the channel signal illustrated in fig. 31.3 is shown in fig. 31.4.

There are separate electron guns in the cathode ray tube of a colour
receiver, one for each of the three primary colours. These guns, although
called the red gun, green gun and blue gun, do of course all emit exactly
the same sort of electron. The inside face of the tube is, however, coated
with special phosphors in a carefully dimensioned pattern so that the
electron beam from the red gun is always focused to hit tiny dots of
material which glow red when hit. The "green electrons" hit spots which
glow green and the "blue electrons" hit spots which glow blue. The
human eye adds these primary colours together so that if, for example, all
three are present on adjacent spots in the correct proportions the area
looks white.

Part G Digital Services

32 **Analog versus Digital**

1 *Two-state Devices*

Most present-day electronic and telecommunication equipment is still analog in nature. This means that the signals to be handled, processed or transmitted are represented by voltages whose amplitude and/or frequency vary continuously with time; thus, in a telephone system, the transmitted signals are replicas of the speech waveforms. Many examples of analog equipment are well known; for example, the radio and television receivers to be found in the majority of homes.

Digital signals are not continuous in nature but consist of discrete pulses of voltage or current which represent the information to be processed. Digital voltages can vary only in discrete steps; normally only two voltage levels are used—one of which is zero—so that two-state devices can be employed. A two-state device is one which has only two stable states; so that it is either ON or it is OFF. Examples of two-state devices are: a lamp which is either glowing visibly or it is not; a buzzer which is either producing an audible sound or not; or an electrical switch which either completes an electrical circuit or breaks it.

The advantages to be gained from the use of digital techniques instead of analog methods arise largely from the use of just the two voltage levels. Digital circuitry, mainly integrated circuitry in modern systems, operates by switching transistors ON and OFF and does not need to produce or to detect precise values of voltage and/or current at particular points in an equipment or system. Because of this it is easier and cheaper to mass-produce digital circuitry. Also, the binary nature of the signals makes it much easier to obtain consistently a required operating performance from a large number of circuits. Digital circuits are generally more reliable than analog circuits because faults will not often occur through variations in performance caused by changing values of components, misaligned coils, and so on. Again, the effects of noise and interference are very much reduced in a digital system since the digital pulses can always be regenerated and made like new whenever their waveshape is becoming distorted to the point where errors are likely. This is not possible in an analog system where the effect of unwanted noise and interference signals is to permanently degrade the signal.

There are two main reasons why the application of digital techniques to both electronics and telecommunications has been fairly limited in scope until recent years. First, digital circuitry was, in the main, not economic until integrated circuits became freely available, and, secondly, the transmission of digital signals requires the provision of circuits with a very wide bandwidth. Some digital circuits and equipments have, of course, been available since pre-integrated circuit days but their scope and application were very limited.

Although digital techniques clearly have significant advantages we live in a world where we usually count in units of 10 and where phenomena such as sound and light are basically analog in nature, i.e. continuously variable waveforms. How then can we convert from a decimal counting system and analog waveforms into this digital world?

2 *Numbering Systems, Coding and Digital Efficiency*

Human beings usually count in tens. No doubt we use a decadic numbering system because we happen to have a total of 10 fingers and thumbs. It is very much simpler, however, to make an electronic device operate at only two levels (On or Off, Yes or No, 1 or 0) than to make it so that it can select any one of ten different levels. This is of course one of the main reasons why digital transmission and switching equipment are now becoming cheaper and more efficient than the analog equivalents.

A two-level numbering scheme is called a BINARY (or base 2) system. Just as our decadic or base 10 system uses zero and figures up to to 9 (i.e. 10 minus 1), a binary or base 2 system uses 0 (zero) and the figure 1 only (i.e. 2 minus 1).

As a simple exercise let us consider the number 1985. In decadic this means

$$(1 \times 10^3) + (9 \times 10^2) + (8 \times 10^1) + (5 \times 10^0)$$

We can change this into straightforward binary in two ways: either on a one-decimal-digit-at-a-time basis called BINARY CODED DECIMAL or taking the number as a whole.

BCD: Binary Coded Decimal

Decimal	Binary Coded Decimal
0	0000
1	0001
2	0010
3	0011
4	0100
5	0101
6	0110
7	0111
8	1000
9	1001

Binary coded decimal for 1985 is therefore

0001   1001   1000   0101

To convert a decimal number as a whole to binary, we divide the decimal number by 2 and then keep on dividing by 2 until there is nothing left. For example,

$$1985 \div 2 = 992 \quad \text{remainder } 1$$
$$992 \div 2 = 496 \quad \text{remainder } 0$$
$$496 \div 2 = 248 \quad \text{remainder } 0$$
$$248 \div 2 = 124 \quad \text{remainder } 0$$
$$124 \div 2 = 62 \quad \text{remainder } 0$$
$$62 \div 2 = 31 \quad \text{remainder } 0$$
$$31 \div 2 = 15 \quad \text{remainder } 1$$
$$15 \div 2 = 7 \quad \text{remainder } 1$$
$$7 \div 2 = 3 \quad \text{remainder } 1$$
$$3 \div 2 = 1 \quad \text{remainder } 1$$
$$1 \div 2 = 0 \quad \text{remainder } 1$$

Therefore, the binary code equivalent of decimal 1985 is

11111000001

i.e. $(1 \times 2^{10}) + (1 \times 2^9) + (1 \times 2^8) + (1 \times 2^7) + (1 \times 2^6) + (1 \times 2^0)$

i.e. Decimal $1024 + 512 + 256 + 128 + 64 + 1$

It will be seen that there are only 11 bits in this binary code for 1985, compared with 16 in the binary coded decimal version of 1985.

Although electronic equipment operates on a binary basis, a Yes–No basis, and we use binary codes for many purposes, computers themselves do not always use straightforward binary for their own calculations. There are several good reasons for this.

Digital efficiency and economics If we use four binary digits (bits) to represent each decimal digit, it is seen from the BCD table that the combinations 1010, 1011, 1100, 1101, 1110 and 1111 have not been used. This means that, since computers have to be designed to handle all the digits input to them, they will be capable of *handling* all 16 of the possible combinations of each 4-bit word but are *using* only 10 of them; they will not be using these 6 combinations at all. In other words, they can at best be only

$$\frac{10}{16} \times 100 \quad \text{or} \quad 62.5\% \text{ efficient}$$

There are basically two common ways round this in order to increase the efficiency of a computer's internal workings:

a) Restricting the machine to only 3 bits per decimal number, i.e. to use an octal or base 8 code, *or*
b) Making good use of all 4 bits per decimal number by using numbers on a Hexadecimal base, i.e. on base 16.

If 1985 is coded on an OCTAL basis we get the following figures:

$$1985 = 3 \times 8^3 \quad (= \text{decimal } 1536)$$
$$+ 7 \times 8^2 \quad (= \text{decimal } 448)$$
$$+ 0 \times 8^1 \quad (= \text{decimal } 0)$$
$$+ 1 \times 8^0 \quad (= \text{decimal } 1)$$

i.e. 3, 7, 0, 1, or in binary coded octal

$$011 \quad 100 \quad 000 \quad 001$$

so a total of 12 bits are needed, compared with 16 bits in ordinary binary coded decimal.

In HEXADECIMAL or base 16 code, the extra numbers are given letter codes, as in the table.

Hexadecimal Code

Decimal	Hexadecimal	Binary Hex
0	0	0000
1	1	0001
2	2	0010
3	3	0011
4	4	0100
5	5	0101
6	6	0110
7	7	0111
8	8	1000
9	9	1001
10	A	1010
11	B	1011
12	C	1100
13	D	1101
14	E	1110
15	F	1111

1985 when coded into hex is therefore

$$(\ 7 \times 16^2) \quad (= \text{decimal } 1792)$$
$$+ (12 \times 16^3) \quad (= \text{decimal} \ \ 192)$$
$$+ (\ 1 \times 16^0) \quad (= \text{decimal} \ \ \ \ \ 1)$$
$$(= \text{decimal } 1985)$$

i.e. 7, 12, 1, or in binary coded hexadecimal

$$0111 \quad 1100 \quad 0001$$

—again using only 12 bits instead of the 16 bits of ordinary binary coded decimal.

To minimize confusion it is not usual for computers to print out answers in octal or hex code; it would be all too easy for mistakes to be made by humans if hex figures were used and mistaken for ordinary decimals: e.g. $707 in hex represents $1799 in ordinary decimals and $707 in octal represents $455.

Some computers use 7-bit words for decimal numbers. This scheme gives each number two parts: the first two bits, the "bi" part, shows whether the number is less than 5 or not:

01 = less than 5
10 = 5 or more

The other 5 bits include only one single 1 code, all the rest are 0s, so the whole representation always has exactly two 1s in it for each digit. If more or less than two are received there has been an error, possibly in transmission, so equipment using this BI-QUINARY code can readily initiate error detection and correction.

Bi-quinary Code

Decimal	Bi-quinary
0	01 00001
1	01 00010
2	01 00100
3	01 01000
4	01 10000
5	10 00001
6	10 00010
7	10 00100
8	10 01000
9	10 10000

3 Analog-to-Digital Conversion

Conversion of analog-type signals such as sound waves to digital format presents more difficulties. The system which is now used for digital telephony was invented in the 1930s by an English engineer, Dr Alec Reeves, working in the ITT Laboratories in Paris. At that time, the technology was not available to make this "pulse code modulation" or PCM system commercially viable for telephony. It has, in fact, taken forty years to reach international agreement on standards to be used for PCM and even now there are two different standards, one basically European, one American. A more detailed description of PCM is given in the next section.

In summary, the speech signal is sampled 8000 times per second; the amplitude of the signal at the instant of sampling is compared with a built-in series of levels called quantizing intervals; the number of the interval within which the sample falls is sent in an 8-digit binary format to the distant end. At the distant end this binary number (representing decimal figures between −128 to +128) is fed into a decoder and a signal with amplitude exactly the same as the original sample is recreated locally. Basically similar sampling and "quantizing" procedures are followed for the conversion of other electrical waveforms into digital signals.

33 Pulse Code Modulation

In a PCM system the analog signal is sampled at regular intervals to produce a pulse amplitude modulated waveform.

If an analog signal is sampled regularly using a sampling rate of at least twice the highest frequency of the signal, the samples are found to be adequate to allow the re-creation of the original voice signal with sufficient accuracy for all purposes. Sampling is done by feeding the analog signal to a circuit with a gate which only opens for the duration of the sampling pulse. The output is a pulse amplitude modulated (PAM) signal (fig. 33.1).

Figs 33.2 and 33.3 show the effect of taking these samples from sound signals of different voice frequencies. Here it has to be remembered that, although the commercial voice band goes up to 3400 Hz, almost all the power of human speech is at much lower frequencies (e.g. at around 500 Hz for male voices), so several samples are taken during each cycle, enough to enable the original analog signal to be reconstituted with a fair degree of accuracy.

The total amplitude range that the signal may occupy is divided into a number of levels, each of which is allocated a number: there are 256 different levels in internationally recommended PCM. Here it should be noted that there are two different PCM encoding laws in use in the world, called μ-law coding (developed and used in America) and A-law coding (developed and used in Europe). These two laws give different quantization values for a signal of a given amplitude so that PCM channels to these two standards cannot directly interwork unless special interfacing equipment is provided.

The variable-amplitude PAM signal is then compared with the instantaneous value of the appropriate range of levels which are called quantizing intervals. The signal is assigned the value of the interval in which it falls and the number of this value is then encoded into an 8-digit binary code. (This gives 2^8 or 256 possible levels, 128 each side of zero.) Each 8-digit code describing the amplitude level of the sample is known as a PCM word. Since the sampling rate is 8000 times per second and each sample has 8 digits, there are 64 000 bits per second for a single PCM channel.

Fig. 33.4 shows how this quantization process works for a complex analog signal. In this figure only 8 sampling levels are shown, for clarity; these 8 levels require only 3 binary digits against the 8 binary digits needed to distinguish between the 256 levels used in practical PCM systems.

The signal waveform is sampled at time instants t_1, t_2, t_3, etc. At time t_1 the instantaneous signal amplitude is in between levels 5 and 6 but since it is nearer to level 6, it is approximated to this level. At instant t_2 the signal voltage is slightly greater than level 6 but is again rounded off to that value. Similarly the sample taken at t_3 is represented by level 2, the t_4 sample by level 2, the t_5 sample by level 1, and so on. The binary pulse train which would be transmitted to represent this signal is shown in fig. 33.5. A space, equal in time duration to one binary pulse, has been left in between each binary number in which synchronization information can be transmitted.

A PCM system transmits signal information in digital form. The quantization process will result in some error at the receiving end of the system when the analog signal is reconstituted. The error appears in the form of quantization noise and can be reduced only by increasing the number of sampling levels. Unfortunately, this would increase the

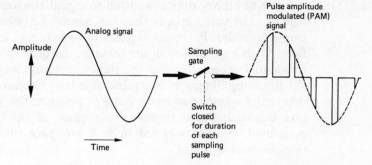

Fig. 33.1 Principle of pulse amplitude modulation (*Courtesy: An Introduction to Digital Telephony, ITT*)

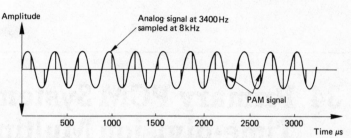

Fig. 33.2 Sampling of the highest voice frequency to be transmitted (*Courtesy: ITT*)

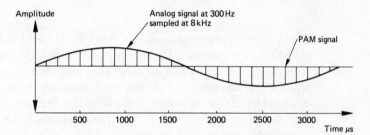

Fig. 33.3 Sampling of the lowest voice frequency to be transmitted (*Courtesy: ITT*)

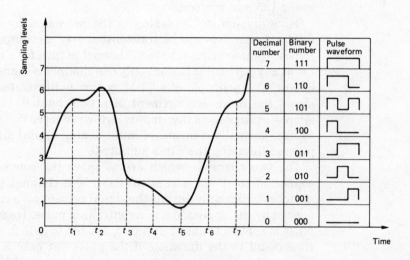

Fig. 33.4 Quantization of a signal

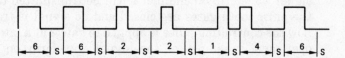

Fig. 33.5 Binary pulse train representing the signal shown in Fig. 33.4

number of binary digits required to signal the sampling level numbers and this, in turn, means that the bandwidth which must be provided would be wider. Practical systems have to accept a compromise solution here. With an 8 binary digits scheme, the quantizing noise is generally considered acceptable. For the receiving equipment to be able to decode the incoming binary pulse trains, it is only necessary for it to be able to determine whether or not a pulse is present. The processes of encoding and quantizing are reversed: a replica of the PAM signal is first generated, then this is fed in to a low pass filter to reconstruct the original analog signal.

34 Primary PCM Systems and Time-division Multiplexing

For a single speech channel an 8-bit PCM word is generated every 125 microseconds (because the sampling rate is 8000 times per second). This complete PCM word can be generated and transmitted so rapidly (in less than 4 microseconds) that any transmission path fed by a single PCM channel would be idle most of the time. By the use of high-speed devices it is possible for PCM words from other channels to be slotted in, to fill this unoccupied time. Each channel is given a designated time-slot, repeated every 125 microseconds.

Time division multiplexing is the procedure by which a number of different channels can be transmitted over a common circuit by allocating the common circuit to each channel in turn for a given period of time, i.e. at any particular instant only one channel is connected to the common circuit. The principle of a TDM system is illustrated by fig. 34.1 which shows the basic arrangement of a two-channel TDM system. In this simple example, analog inputs are considered for clarity. TDM is, of course, normally a means of multiplexing digital signals, such as PCM pulses all having the same amplitude.

The two channels which are to share the common circuit are each connected to it via a channel gate. The channel gates are electronic switches which only permit the signal present on a channel to pass when opened by the application of a controlling pulse. Hence, if the controlling pulse is applied to gate 1 at time t_1 and not to gate 2, gate 1 will open for a time equal to the duration of the pulse but gate 2 will remain closed. During this time, therefore, a pulse or sample of the amplitude of the signal waveform on channel 1 will be transmitted to line. At the end of the pulse, both gates are closed and no signal is transmitted to line. If now the controlling pulse is applied to gate 2 at a later time t_2, gate 2 will open and a sample of the signal waveform on channel 2 will be

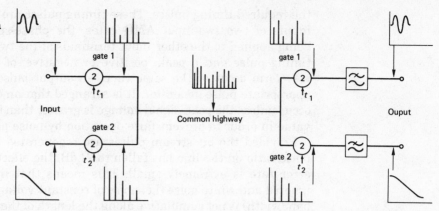

Fig. 34.1 A simple two-channel TDM system: t_1 = a series of pulses occurring at fixed intervals; t_2 = a series of pulses occurring at the same periodicity as t_1 but commencing later by an amount equal to half the time interval

transmitted. Thus if the pulses applied to control the opening and shutting of gates 1 and 2 are repeated at regular intervals, a series of samples of the signal waveforms existing on the two channels will be transmitted.

At the receiving end of the system, gates 1 and 2 are opened, by the application of control pulses, at those instants when the incoming waveform samples appropriate to their channel are being received. This requirement demands accurate **synchronization** between the controlling pulses applied to the gates 1, and also between the controlling pulses applied to the gates 2. If the time taken for signals to travel over the common circuit was zero, then the system would require controlling pulses in exact synchronization at both ends, but since, in practice, the transmission time is not zero, the controlling pulses applied at the receiving end of the system must occur slightly later than the corresponding controlling pulses at the sending end. If synchronization signals are sent from one end to the other as an integral component of the PCM system (as they are with internationally specified systems), all the signals will of course maintain their correct relative positions. If the pulse synchronization is correct, the waveform samples are directed to the correct channels at the receiving end. The received samples must then be converted back to the original waveform, i.e. demodulated.

In its passage along a telephone line, the TDM signal is both attenuated and distorted but, provided the receiving equipment is able to determine whether a pulse is present or absent at any particular instant, no errors are introduced. To keep the pulse waveform within the accuracy required, pulse regenerators are fitted at intervals along the length of the line. The function of a pulse regenerator is to check the incoming pulse train at accurately timed intervals for the presence or absence of a pulse. Each time a pulse is detected, a new undistorted pulse is transmitted to line and, each time no pulse is detected, a pulse is not sent.

The simplified block diagram of a pulse regenerator is shown in fig. 34.2 The incoming bit stream is first equalized and then amplified to reduce the effects of line attenuation and group-delay/frequency distortion. The amplified signal is applied to a timing circuit which generates the required timing pulses. These timing pulses are applied to one of the inputs of two two-input AND gates, the phase-split amplified signal being applied to the other input terminals of the two gates. Whenever a timing pulse *and* a peak, positive or negative, of the incoming signal waveform occur at the same time, an output pulse is produced by the appropriate pulse generator. It is arranged that an output pulse will not occur unless the peak signal voltage is greater than some pre-determined value in order to prevent false operation by noise peaks.

Provided the bit stream pulses are regenerated before the signal-to-noise ratio on the line has fallen to 21 dB, the effect of line noise on the error rate is extremely small. This means that impulse noise can be ignored and white noise (i.e. noise of constant voltage over the operating bandwidth) is not cumulative along the length of the system. This feature is in a marked contrast with an analog system in which the signal-to-noise ratio must always progressively worsen towards the end of the system. Thus, the use of pulse regenerators allows very nearly distortion-free and noise-free transmission, regardless of the route taken by the circuit or its length (fig. 34.3).

There are, as pointed out in the previous section, two different types of PCM, one basically European, the other American in origin. Different quantizing laws are used, together with different signalling and synchronization procedures and a different higher-level multiplexing structure:

"European" (A-law) PCM

Primary multiplexing	30 channels	2.048 Mbit/s
2nd order	120 channels	8.448 Mbit/s
3rd order	480 channels	34.368 Mbit/s
4th order	1920 channels	139.264 Mbit/s
5th order	7680 channels	560.000 Mbit/s

"American" (μ-law) PCM

Primary multiplexing	24 channels	1.544 Mbit/s
2nd order	96 channels	6.312 Mbit/s
3rd order	672 channels	44.736 Mbit/s
4th order	4032 channels	274.176 Mbit/s

Modern PCM systems terminals make much use of integrated circuits, many functions being performed in a single chip. To enable principles to be shown, a schematic diagram of an early type of 30-channel PCM is given in fig. 34.4. It will be seen that the 2048 kbit/s line signal is produced in stages:

1) Analog input is fed through an 8 Hz gate, giving in effect PAM or pulse amplitude modulated samples.
2) These PAM samples are multiplexed together on a time-division basis.
3) The multiplexed PAM samples are encoded; this quantizes the levels of the samples and prepares the strings of binary digits to indicate these amplitude levels.
4) Synchronization is established.
5) Frames are aligned and signalling inserted.

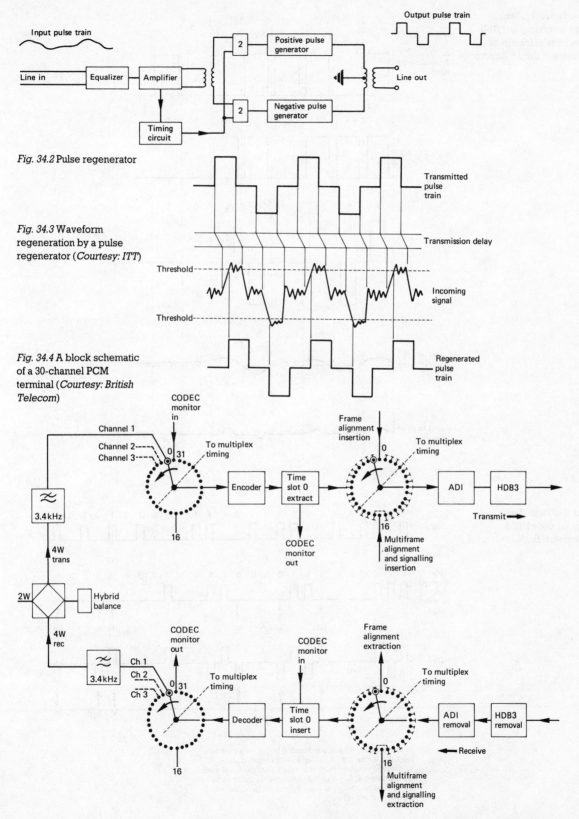

Fig. 34.2 Pulse regenerator

Fig. 34.3 Waveform regeneration by a pulse regenerator (*Courtesy: ITT*)

Fig. 34.4 A block schematic of a 30-channel PCM terminal (*Courtesy: British Telecom*)

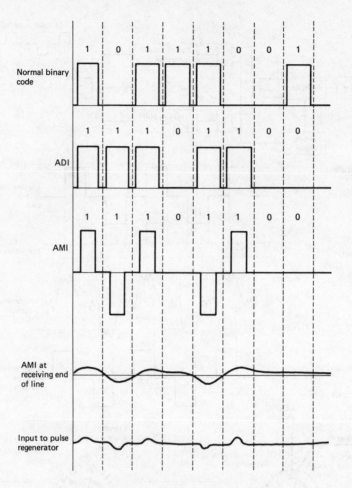

Fig. 34.5 ADI (alternate digit inversion) and AMI (alternate mark inversion) (*Courtesy: British Telecom*)

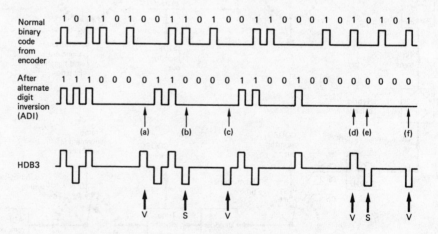

Fig. 34.6 High-density bipolar code HDB3 (*Courtesy: British Telecom*)

In the example, 4 zeros are detected by (a), so a violation mark (V) is inserted. Another 4 zeros are detected by (c), so a second violation mark is inserted. An additional mark (S) is inserted at (b). Further violations are detected by (d) and (f) and an additional mark in inserted at (e) but not between (c) and (d).

6) The binary signals representing the samples go through an ADI or Alternate Digit Inversion stage (see fig. 34.5) to overcome some of the disadvantages of unipolar transmission.

7) These ADI signals then go through an AMI or Alternate Mark Inversion stage in order to produce a signal which may be transmitted and regenerated with minimum distortion.

A modified form of AMI called HDB3 (High-Density Bipolar 3) is nowadays used. This overcomes difficulties caused by the fact that ordinary AMI does not code zeros in any way, so if there is a long sequence of consecutive zeros it is difficult to maintain the correct timing relationship between the receive terminal and the received signal. With HDB3, after three successive zeros, the fourth binary zero is replaced by a mark signal of the same polarity as the previous mark signal (see fig. 34.6). Because this is a violation of the alternate mark inversion rules, the receiving equipment knows that this received mark signal is not a genuine mark but represents another zero. If either an even number of marks or no marks at all exist between each violation, then an additional mark (S in fig. 34.6) is substituted in place of the first of the four zeros The substitution mark has a polarity opposite to that of the previous mark.

35 Data Services

1 Computer-based Data Links
The coming of the microprocessor means that many offices which could not economically justify the purchase of a large main-frame computer are now able to increase their efficiency by using digital computers to carry out many tasks which were previously expensive and labour intensive, such as payroll preparation.

There is still however a huge demand for interworking between computers, minicomputers and microcomputers at different locations. The high-street banks, for example, make use of computers to maintain details of customers accounts, of standing orders, direct debits, etc., while airlines and package holiday firms are able to operate booking systems that provide an immediate confirmation of vacancies and bookings. Fig. 35.1 shows the world-wide airline bookings system operated by SITA (Societe Internationale de Telecommunications Aeronautiques), the Paris-based international body which co-ordinates such airline activities. This has a "high-level" network of 10 main switching nodes linked together by 9.6 kbit/sec circuits working on a packet-switched basis. These work to a second level of computerized nodes and 48 satellite processors, and these in turn work down to more than 5000 terminals, all over the world.

The cost of a digital computer with a large storage capacity is high and so it is not economic for an organization to instal a computer at all the points in its offices and factories where a computing facility would be of use. It may, however, be economic for a main-frame computer to be installed only at one, or perhaps two, points in the organization's network of offices. For the branch offices to have on-line access to computing facilities it is necessary for them to be connected to the computer centre by means of data links.

Organizations sometimes find that it is attractive for various reasons to be able to share access to their computers and databases with other parts of their own local organization. To enable one computer to inter-work satisfactorily with others in the same area, Local Area Networks, or LANS, have been developed. Some of these specialized LANS are sometimes called Rings (see Section 38).

Many smaller businesses are unable to economically justify the cost of purchasing and operating a main-frame computer and yet have a need for computing facilities. To meet this demand computer bureaux have been set up which rent computer services to their customers on a time-sharing basis. Customers need data links to these bureaux.

2 Privately Leased Circuits and the PSTN

A data link that connects a data terminal to a remote digital computer may be leased from the telephone administration or it may be temporarily set up by dialling a connection via the public switched telephone network (PSTN). The choice between leasing a private circuit and using the PSTN must be made after the careful consideration of factors such as the cost, availability, speed of working, and transmission performance. Private circuits may transmit d.c. ± 6 V or ± 80 V signals or may use modems to convert the data signals into voice-frequency signals. DC can only be used if there is a metallic path all the way; if the circuit is provided by any form of multiplexing, either FDM or TDM, the signals have to be fed through a modem. Voice-frequency data circuits can also be of any length and may be routed, wholly or partly, over loaded or unloaded audio-frequency cable, or over multi-channel telephony systems. A modem (short for modulator–demodulator) converts digital signals from a data terminal into modulated analog signals which can be carried by the PSTN. These analog signals carrying digital information are sometimes called pseudo-digital signals.

In many countries the cost of permanently leasing a line may be relatively high and it could only be economically justified if there is sufficient data traffic on the line or the particular terminal application necessitates a permanent connection, e.g. a cash dispensing terminal in a bank which checks a customer's account before releasing the cash. If the data communication requirements involve occasional contact with a large number of locations, and the majority of the connections are of fairly short duration, the use of the PSTN is probably best. On the other hand if long-duration connections between a few branch offices and the computer centre are likely, leased links will probably be chosen. In

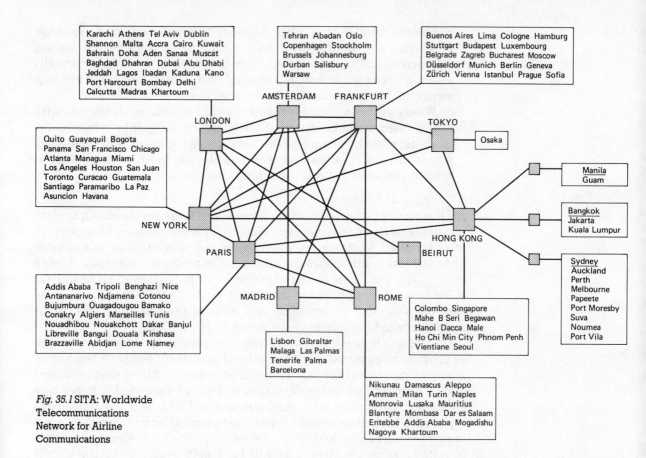

Karachi Athens Tel Aviv Dublin
Shannon Malta Accra Cairo Kuwait
Bahrain Doha Aden Sanaa Muscat
Baghdad Dhahran Dubai Abu Dhabi
Jeddah Lagos Ibadan Kaduna Kano
Port Harcourt Bombay Delhi
Calcutta Madras Khartoum

Tehran Abadan Oslo
Copenhagen Stockholm
Brussels Johannesburg
Durban Salisbury
Warsaw

Buenos Aires Lima Cologne Hamburg
Stuttgart Budapest Luxembourg
Belgrade Zagreb Bucharest Moscow
Düsseldorf Munich Berlin Geneva
Zürich Vienna Istanbul Prague Sofia

AMSTERDAM   FRANKFURT

LONDON   TOKYO

Osaka

Quito Guayaquil Bogota
Panama San Francisco Chicago
Atlanta Managua Miami
Los Angeles Houston San Juan
Toronto Curacao Guatemala
Santiago Paramaribo La Paz
Asuncion Havana

Manila
Guam

NEW YORK

Bangkok
Jakarta
Kuala Lumpur

HONG KONG

PARIS   BEIRUT

Sydney
Auckland
Perth
Melbourne
Papeete
Port Moresby
Suva
Noumea
Port Vila

Addis Ababa Tripoli Benghazi Nice
Antananarivo Ndjamena Cotonou
Bujumbura Ouagadougou Bamako
Conakry Algiers Marseilles Tunis
Nouadhibou Nouakchott Dakar Banjul
Libreville Bangui Douala Kinshasa
Brazzaville Abidjan Lome Niamey

MADRID   ROME

Colombo Singapore
Mahe B Seri Begawan
Hanoi Dacca Male
Ho Chi Min City Phnom Penh
Vientiane Seoul

Lisbon Gibraltar
Malaga Las Palmas
Tenerife Palma
Barcelona

Nikunau Damascus Aleppo
Amman Milan Turin Naples
Monrovia Lusaka Mauritius
Blantyre Mombasa Dar es Salaam
Entebbe Addis Ababa Mogadishu
Nagoya Khartoum

Fig. 35.1 SITA: Worldwide
Telecommunications
Network for Airline
Communications

practice, most private data networks consist of a combination of both
leased and PSTN links, and very often the leased circuits are provided
with the stand-by facility of using the PSTN when necessary (i.e. if the
leased circuit should fail).

3 Switched Circuits

The use of the PSTN to carry data traffic often restricts working speed
significantly, and call set-up time limits interactive working between
terminals. In recent years many telecommunications administrations
have recognized that there is a genuine public demand for improved data
facilities, and are now beginning to provide public switched data networks so that customers are spared the expense of leasing long-distance
circuits and avoid the operating constraints of using the PSTN.

There are three basic ways of switching record traffic or data:

a) *Circuit switching:* a complete communications circuit is established
for the exclusive real-time use of the subscribers for the duration of the
call.

b) *Message switching:* this is a store-and-forward technique which is
used for message traffic but is not appropriate for all data purposes.
The various types of magnetic or semi-conductor stores now used are a
great improvement over the "tornpaper tape" stores which were

commonly used in the 1970s. Even so, message switching can sometimes run into significant delays of minutes or even hours, making interactive traffic impossible. Since interactive working is normally essential between a remote terminal and a host computer, message switching is not practicable for such services.

c) *Packet switching:* data streams are split up into short packets with well-defined formats. Because there are no long data streams, the blocking of links is rare in well-engineered systems and end-to-end delays are small, often less than one second. Interactive working is practicable, as with circuit switching. (See Section 36.)

4 *Telex and Teletex*

At the present time, far more firms have telex machines than data terminals working to computers. Many improvements have been made in telex terminal equipment in recent years, e.g. machines are now quieter and need less maintenance, messages can be prepared on a visual display unit (VDU) and then stored electrically; it is no longer necessary for coded holes to be punched in long paper tapes. Telex exchanges also are now often computer controlled and able to provide such services as instant printed out advice of the duration of each call.

Despite all these improvements the telex service must now be regarded as about to be superseded—international agreement has been reached on a new service called Teletex. This will provide everything that telex does and do it faster and better. Messages will not be restricted to upper case characters as on most telex machines but will look as if they have been typed on a high-class electric typewriter—capital letters, margins, spacings, type variations and so on. Teletex services have already begun to operate in some countries and will be widely available by the middle 1980s. These services will depend on the availability of circuits able to carry digital signals at 2.4 or 4.8 kbits/s with no significant distortion.

5 *Word Processors*

Many companies now use word processors instead of typewriters; they are especially valuable when many copies of generally similar documents have to be prepared or when amendments continually have to be made to a prepared text. Most word processors are at present used by staff at a single location; it is not often economic to arrange for a word processor in one town to be accessed by staff in another.

The coming generation of word processors is however likely to be different; they will be Communicating Word Processors (CWP). The protocols under international discussion for teletex services are likely to be the basis on which most CWPs will work. A letter prepared and checked on a VDU in one town will, on the touch of a key, be transmitted digitally to and printed out at another town. A few seconds for the transmission and printing of a complete A4 page of typing will soon become the norm.

36 Packet Switching

In the ordinary circuit-switched Public Switched Telephone Network (PSTN) an end-to-end analog service is provided by setting up dedicated paths comprising fixed point-to-point circuits joined in temporary association by switched contacts at exchanges. Path set-up is controlled by the calling subscriber (see fig. 36.1). If we use a PSTN circuit to carry data, we need a modem (modulator plus demodulator) at each end of the circuit

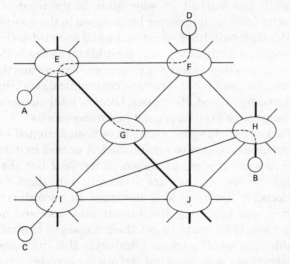

Fig. 36.1 Circuit switched network: calls from A to B and from C to D are established by paths interconnected at switching centres E, F, G, H, I

to convert the digital signals of our data terminal to an analog form suitable for transmission over the PSTN. A good modem and a good circuit will permit a bit-rate of 4.8 kbit/s but on most circuits in most countries 2.4 kbit/s is all that can be managed without too many errors being introduced. It takes time to set up a PSTN call, often several seconds, and as the network gets busier the call-set-up time increases until there is congestion, and no calls can be established at all even if the wanted subscriber at the distant end is free and waiting for your call. Most telephone calls last for several minutes and in many countries you pay for calls in units, 1 unit call for every 3 minutes or less, so a few seconds for the call to be set up is acceptable. But data calls sometimes need last for only a fraction of a second and it would be very wasteful if you had to pay for 3 minute's use of a circuit which you had in fact used for, say, only one-tenth of a second!

Another way of switching data traffic is to use message switching, or "store and forward" switching (see fig. 36.2). Message switching is similar to circuit switching in that the network comprises a number of switches interconnected by point-to-point circuits, but, additionally, at each switching node there is a memory store into which messages may be passed for temporary storage or buffering. At each such node, the destination address of each message is examined to determine whether the message

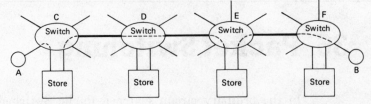

Fig. 36.2 Message switching: store and forward: message from A to B is stored temporarily at C and E because circuits forward are busy. As soon as circuit forward is free, the message is pumped from the store on towards its destination

is to be transmitted on to another node or if it is to a destination terminated on that switch itself. If the outgoing link, or the circuit to the destination address, is free, the message is sent straight on. If the required link or destination is busy, the message is stored or buffered in a queue of messages in the local memory. Transmission of the message is tried again when it has worked its way again to the front of the queue; different priority levels can however be assigned in the queue.

Message switching using store-and-forward techniques does therefore introduce a finite delay, which builds up as the buffer stores in each node fill with messages awaiting clearance. This means that message switching cannot be used for interactive communications, that is to say messages demanding immediate replies, because total queueing time can sometimes be minutes or even hours, not just nanoseconds.

Packet switching networks have been designed so that interactive data communication *can* be established. A packet switched network is a store-and-forward process as shown in fig. 36.2 but the messages are not of indefinite length, they are divided up into short fixed lengths known as packets. A few very long messages in an ordinary message switching system can hold long-distance circuits busy and make it impossible for other would-be users to get their messages transmitted. The division of traffic into small packets eliminates this difficulty. In a well designed system there is no blocking and queues are very small.

Each packet carries in its "header" (the first few digits) the address to which it is to be forwarded and, as with ordinary message switching, these "address to" digits are automatically analysed at each switching node and the packet either sent straight on its way or (if all onward circuits are busy or out-of-order) put into the local temporary store. This means that, on the main highways between switching nodes in the network, the packets received from one terminal are likely to be interleaved with packets from many other terminals. Because the "address to" digits are read and acted upon at each node, some packets may be sent one way and some on other routes to the same final destination. At the destination, the packets into which a complete message was divided are automatically re-assembled into their correct order so that the message becomes a sensible whole.

A packet switching network can be used by "non-intelligent" data terminals like teletype machines as well as by specially designed packet-mode terminals. The signals received from a teletype or similar machine are dealt with by a PAD (Packet Assembler-Disassembler) which takes in signals character-by-character and builds them up into standard packets which it sends out to the packet network (see fig. 36.3). Because the network uses store-and-forward procedures it can be used by terminals using many different data rates. If a message from a fast terminal is

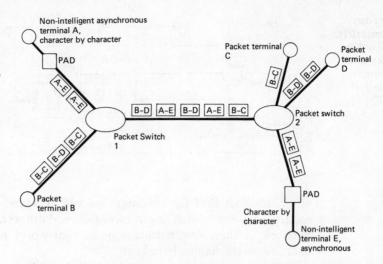

Fig. 36.3 Packet switching network with packet assembler-disassembler (PAD) to interconnect non-intelligent teletypewriter to the network, showing packets interleaved on the link between nodes. PAD functions are detailed in CCITT Rec.X.3, and interworking with PADs is described in X.28 and X.29

to be delivered to a slow terminal, the message merely stays in buffer stores a bit longer, because the last link out to the destination terminal is restricted to the slower speed of that terminal.

Each packet also has a trailer with an "error checksum", used for error correction. This error checksum ensures that the packet has been received without errors. Before the days of integrated circuits this procedure would probably have needed a whole mainframe computer of its own. The information in the packet, excluding the "flags" which indicate the beginning and ends of signals, is considered for this checking purpose as being a string of numbers, all in binary code. The total figure is divided by a 16-bit number (internationally standardized), the quotient is disregarded, and the remainder added to the packet to be transmitted as a separate checksum. At the receiving end, the whole packet including this extra checksum is again divided by the same 16-bit number. It should divide exactly, with nothing left over. If there is a remainder, it means something has gone wrong during transmission so the whole packet is re-transmitted, automatically, until it *is* received correctly.

Packet switching is simple in its basic essentials as given above but extremely complicated in all its details, especially as these are all of international validity.

A few definitions will however be of interest:

A *datagram* is a packet which is a self-contained message; it is routed individually through the packet network rather like a telegram going through a message switching system.

A *virtual call* is a bit more difficult. If one terminal in a packet network wants to send a long message to another terminal, it begins by telling the distant-end node that a long message is coming and special indicating signals called logical channels are established. From then on all the many packets that make up the complete message are sent through the network just like other individual packets but at the destination end they are immediately assembled in the correct order and delivered to the destination terminal as one complete message. For all practical purposes, the two users have a direct private circuit from one terminal through to the

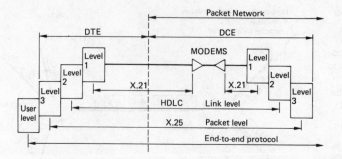

Fig. 36.4 A user's data terminal equipment (DTE) and its interface with the packet network whose entry point is the data circuit-terminating equipment (DCE)

other; the fact that the message has been divided up into hundreds of separate packets which might have followed different routes through the network to their destination is of no real worry or importance to the user, hence the name virtual call.

Protocol means the rules and regulations. For packet switching, the protocols are given in CCITT Recommendation X.25.

Layered Protocol For data work it is very convenient to divide up the various activities concerned into functional layers. For example, if you are using an ordinary telephone you do not need to understand how the microphone turns your voice into electrical signals, and the microphone does not need to know how the number you dial connects you with the friend to whom you wish to speak: so long as each "layer" deals satisfactorily with its own particular level of responsibility all will be well. The International Organization for Standardization (ISO) has developed a layered architecture to help data users all over the world inter-work their data systems: the model is called OSI, for Open System Interconnection, and its layers will be defined in the next paragraph.

The ISO is for business machines and computers very roughly what the CCITT is for telecommunications—it defines standards which make for easier interworking. Telecommunications and Computers are now very closely linked together indeed—packet switching involves both the CCITT and the ISO.

The OSI layers are

Layer 1 – *Physical:* the actual electrical and mechanical physical components of the circuit.

Layer 2 – *Link:* the circuitry which enables data to be moved between nodes directly connected together.

Layer 3 – *Network:* the equipment which provides the interface between nonpacketized data and the packetized system; this layer packetizes and re-assembles messages and controls the flow of packets in the network.

Layer 4 – *Transport:* this is the layer which moves the message right across from origin to destination.

Layer 5 – *Session:* this layer establishes, maintains and ultimately terminates end-to-end logical links.

Layer 6 – *Presentation:* this layer edits, maps and translates data into an appropriate form; it provides an acceptable common language.

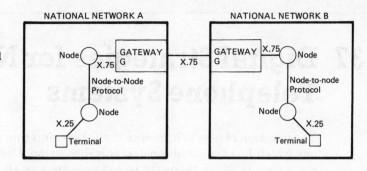

Fig. 36.5 International packet working using gateway exchange G, showing the protocols and specifications for each section

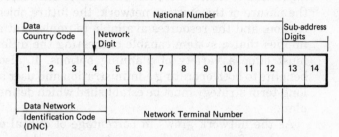

Fig. 36.6 The internationally agreed numbering plan for packet switched networks (CCITT Rec.X.121)

Layer 7 – *Application:* this is the actual conversation between the data terminals concerned.

CCITT Recommendation X.25 defines the interface between a user's packet-mode terminal (Data Terminal Equipment, DTE) and the packet network whose entry point is called a Data Circuit-terminating Equipment, DCE. Fig. 36.4 shows the relationship between DTE and DCE. X.25 also defines three basic levels, in line with the layers defined in OSI:

X.25 Level 1: provides a synchronous, bit-serial full-duplex point-to-point circuit for the transmission of data between user's equipment and the network (Rec X.21 defines the interfaces).

X.25 Level 2: is the link control or frame level, it deals with the detection of transmission errors and their correction. Another ISO specification is used here, that for High-level Data Link Control or HDLC, to provide the error-free transmission system.

X.25 Level 3: defines three basic types of packet service:
    a) a single packet service, the datagram
    b) a multi-packet service provided on a switched basis
    c) a multi-packet service provided on a permanent basis.

International interworking between packet-switched data networks is now possible; there are now many such networks all over the world. Fig. 36.5 shows how a terminal in one country can access a network in another country, using their International Gateway Exchanges, G on the diagram. An internationally agreed numbering plan has to be used; this is given in detail in CCITT Recommendation X.121, and is shown in Fig. 36.6.

37 Digital Strategies for National Telephone Systems

The question of how to introduce digital systems into existing networks is one which has provoked considerable discussion. Unfortunately, it is not a question to which there can be a simple answer, since much depends on the nature of the existing network, the future objectives of the administration, and the resources available to implement any programme. This implies that a system capable of meeting the different needs that arise must offer a variety of planning options. It also implies that if the benefits to both operating administration and user are to be maximized, a long-term strategy must be established which defines a complete network plan.

As the network grows in percentage of digital operation, the longer-term benefits begin to become apparent to the customer. Transmission performance of integrated digital networks offers greater consistency between different types of call and will give lower overall loss. Availability of end-to-end digital connections then allows the introduction of new digital services by provision of digital operation over the local line plant to the customers terminal, finally arriving at the integrated services digital network.

There are four basic strategies which can be identified for introducing new systems such as digital exchanges:

1) *Augmentation* where an existing network mode is extended or a new one added with minimum alteration of the overall network topology.

2) *Replacement* where obsolete equipment has reached the end of its economic service life.

3) *The digital island approach:* isolated cell growth where a group of exchanges can be considered together for modernization (fig. 37.1).

4) *Overlay* where a complete network coverage of new equipment is provided (fig. 37.2).

The augmentation method has limitations in terms of the amount of interworking requirements, possibly leading to economic penalties, whilst offering little improvement in service or performance due to the interworking constraints.

The replacement and digital island methods can offer more immediate user benefits, at least in limited areas, whilst the overlay method offers more rapid and widespread user and administration benefits.

The criteria which are used in planning in many countries are as follows. If several local exchanges in an area are to be augmented or replaced, then an overlay toll or tandem facility is provided. All further growth should be taken on the new equipment to minimize interworking requirements. Traffic originating in the new network should remain in that network as long as possible to minimize the number of new/old or old/new transitions experienced by calls.

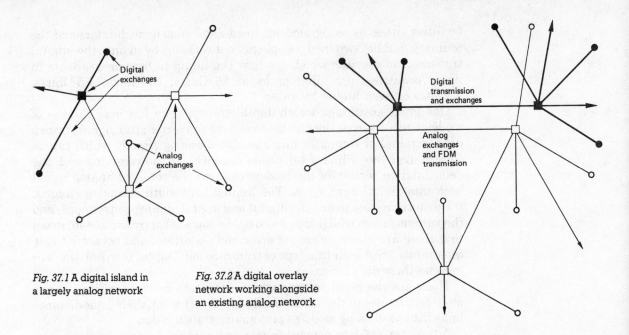

Fig. 37.1 A digital island in a largely analog network

Fig. 37.2 A digital overlay network working alongside an existing analog network

The progressive evolution of a new network of the overlay type can provide a good opportunity for rationalizing signalling plans. Existing networks tend to grow with a variety of equipment and signalling types, giving problems in maintenance, training and spares holding. The introduction of common channel signalling between the new elements of the network particularly by taking advantage of the quasi-associated signalling options can lead to the establishment of a coherent signalling network and a further limitation of interworking requirements.

38 Local Area Networks (LANs)

1 *Area Networks (LAN, WAN, MAN)*

LANs have been defined as networks permitting the interconnection and intercommunication of a group of computers, primarily for the sharing of resources such as data storage devices and printers. LANs cover short distances (less than 1 km, usually), almost always within a single building complex. Different data transfer rates are possible and shared centralised data storage access is available. Response times are comparable with those of a single computer.

Networks which have been designed to carry data calls over long distances (many hundreds of kilometres) also exist; these are usually called WANs, or Wide Area Networks. WANs provide their long-distance facilities either by using modems (modulator plus demodulator) and the ordinary public switched telephone network or by using the digital transmission services which are now beginning to become available in many countries, providing paths at 56 Kbit/s, 64 Kbit/s, 1.5 Mbit/s, 2 Mbit/s or even higher bit rates.

One great advantage which digital transmission has over the use of modems results from the fact that modems convert digital signals from a data terminal or computer into modulated analog signals which can be carried over the ordinary telephone network. At the receiving end, the demodulation section of the modem changes the received analog signal back into digital form again. The limited bandwidth of analog circuits, the distortions due to analog/digital and digital/analog conversions, and the various noises which are picked up during analog transmission, mean that there are many sources of error and distortion, and severe bit-rate constraints arise with this type of transmission. Digital transmission eliminates these difficulties.

These are the main reasons why many users are now willing to pay more for the use of digital transmission systems for their long-distance links instead of using modems plus analog transmission.

A third type of data network is now beginning to be developed in major cities, with ranges between the purely local LAN and the long-distance WAN. These are called Metropolitan Area Networks, or MANs. MANs are likely to become common and extremely important as cable TV systems grow, and they could well develop into backbone networks used to carry digital transmission systems right across our cities at high bit rate and with minimum distortion.

2 Transmission: Broadband or Baseband

Two different transmission practices can be used in LANs, WANs and MANs; each has advantages in different circumstances.

Broadband systems are networks in which very wide bandwidths (high bit rates) are used. This bandwidth is often either permanently subdivided into several separate data channels using frequency division multiplexing (FDM) techniques, or "frequency agile" modems are used which listen out on the network, find a free data channel, and then transmit out on this channel. Very few data users at present really need high bit rates (i.e. 10 to 20 Mbit/s). Tremendous amounts of information need to be carried to justify the expense of using such speeds—but to transmit video signals, particularly moving pictures of TV standard, then bit rates of 70 Mbit/s or more are usually needed. Broadband systems are therefore beginning to be talked about as the cable TV systems of the future, providing a wide choice of good-quality TV channels, together with interactive data and voice (telephony) services.

Baseband services are networks in which all the digital signals are carried directly on the cable. Brief descriptions of the various types of network currently used in baseband systems are given in the following paragraphs.

3 Network Topologies and Interconnection Technologies

If you are going to connect together many different transmitting and receiving devices, there have to be ways of ensuring that the various transmissions all go to the desired receiver without being mixed up with other signals. There are several basic methods used:

a) *Bus Polling* All users are connected to a single circuit, called a bus. A controlling device regularly scans (polls) the bus to find out if any of the equipments connected to it want to transmit messages. If there is a message awaiting transmission, the access controller authorises that particular station to take over the bus. Then as soon as the message has been sent, the access controller takes back control and carries on polling, to give the other terminals an opportunity of sending their traffic. (See fig. 38.1.)

Hi Net is a typical Bus Polling System.

b) *Bus Contention* Here again all users are connected to a single bus; each terminal has to listen out on the bus for itself, and can only transmit

Fig. 38.1 A bus system

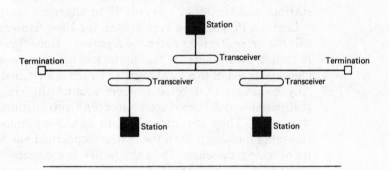

its own outgoing messages when it finds that the bus is free, with no one else transmitting. The circuitry which watches out for a free gap in this way is called Carrier Sense Multiple Access or CSMA.

If two terminals, both anxious to transmit their messages, start transmitting simultaneously, this is called a collision. There are several ways of eliminating collision difficulties (sometimes called double seizure or contention). One of the most common requires the station doing the transmitting also to listen out to the received signals on the bus and to compare those with the signals it is sending out. If some other station happens to have spotted (and seized) the same gap and is also transmitting, then the received signals will not be exactly the same as those being transmitted. (They will be mutilated and corrupted by the collision.) Transmission is stopped, and the terminal tries again later. This is called Collision Detection, CD: the totality of the system is called CSMA/CD.

Ethernet is a typical CSMA/CD system.

c) *Ring Systems* Here all users are joined to a ring. Each message goes right round the ring and back to its source for integrity checking, having been received by each station on the ring and regenerated before being sent on round.

In the basic or "Empty-Slot" type of ring systems, a controlling station called the monitor periodically sends out short packets of signals into the ring, indicating that these particular time slots are empty and available for use. If any station wants to transmit a message, it substitutes that message for the "empty" signal bits generated by the monitor. The transmitted message is detected and taken out of the ring when it has reached its destination station.

The Cambridge Ring is a typical empty-slot ring system (see fig. 38.2).

Various modifications have been suggested to basic ring networks in order to improve performance, particularly under fault condition. Examples of these are:

(i) *Dual Ring Systems* (fig. 38.3)   Two complete paths are provided round the ring on a main-plus-standby basis. If the main path fails in any section, the standby path is switched in automatically, to keep the ring complete.

(ii) *Braided Ring Systems* (fig. 38.4)   Three separate paths are, in effect. provided round the ring, with one of these paths fed in to all of the stations and the other two only in to alternate stations.

Logica's Polynet is a typical Braided Ring System.

(iii) *Buffer or Register Insertion Systems*   Here there is a single ring but it is not continuous. When a station has a message to transmit, the message is fed in to a local register which is switched to the outgoing line. Any messages that come in from around the ring addressed to other stations are not then regenerated and retransmitted automatically at that point. They are instead fed in to a delay buffer which holds these incoming messages until the station concerned has finished transmitting its outgoing message. Then the buffer is connected to the outgoing line and the messages stored in it follow the new message round the ring. Incoming messages addressed to a station carry an appropriate label, and when these arrive they are fed straight into the local receive buffer and in to the local terminal.

Hasler's SILK system is a typical Register Insertion System.

4 Token Systems

Tokens are another way of avoiding double seizure collisions: a special flag or token signal gives authority to a terminal to begin transmitting a signal in to the network. There is only one of these tokens; it is not available for a second terminal until the transmission from the first has been completed. The token is then sent around the ring and is available to be seized by another terminal, giving it the authority to transmit. (Token working could be called a digital form of pass-the-thimble).

In a star-ring token system (fig. 38.5), each terminal is connected to the ring via two paths, one incoming, the other outgoing. Normally the complete ring circuit goes through each of the terminals via their star connections. If there is a fault on the star or at the terminal itself, the ring is put straight through, and the star circuits to and from the terminal are looped together so that the terminal may test itself and its star link before rejoining the network.

The IBM token ring is a typical star-ring system.

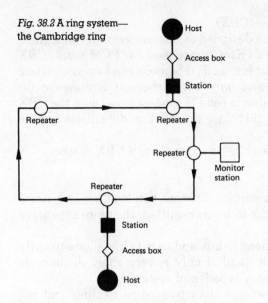

Fig. 38.2 A ring system—
the Cambridge ring

Host

Access box

Station

Repeater

Repeater

Repeater

Monitor
station

Repeater

Station

Access box

Host

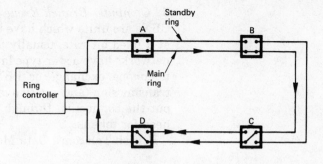

Fig. 38.3 A dual ring
system. Failure between C
and D results in the standby
ring circuit being switched
in for use between those
two points

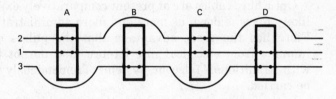

Fig. 38.4 A braided ring
system. As with the
ordinary dual ring system,
two circuits are available in
each direction from each
station. "Braiding" (and the
existence of the third
circuit) reduces the
likelihood of fault

Fig. 38.5 A star-ring token
system

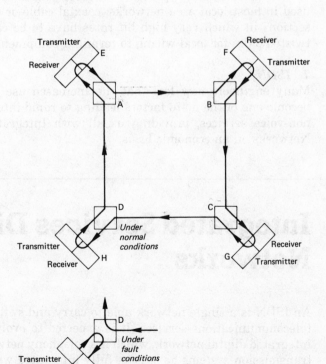

5 *Computer Branch Exchanges (CBX)*

CBXs are units which have been designed to interconnect digital signals at a fixed bit rate, usually the 64 Kbit/s rate used for PCM voice. CBX networks have a star-type layout just as do telephone lines connected to a telephone exchange or PBX, and only when there is a message for transmission (and a path available to send the message on) does the CBX put the connection through so there are no collision difficulties and no need for tokens.

British Telecom's Data Monarch is an example of a CBX system.

6 *Types of Cable*

Different transmission systems require different bandwidths: in general the higher the bit-rate which has to be transmitted, the more expensive the bearer.

Simple twisted-pair cables, cheap to buy and easy to instal, are usually adequate for baseband systems—and if only a very short distance is involved, can also be used to carry broadband systems.

Coaxial cables are more expensive than twisted-pair cables and not quite as flexible but they can carry much wider bandwidths.

Optic fibre cables are at present comparatively expensive but prices are likely to come down as more and more administrations go over to fibre. Fibre has the great advantage that the pulses of light utilised are immune from electrical interference, and do not themselves interfere with "traditional" telecoms systems. Tremendously wide bandwidths can be carried.

In the real world it is probable that a combination of cable types will be used in most local area networks: coaxial cable or optic fibres for those sections in which very high bit rates have to be carried, with cheaper twisted pairs for local wiring to terminal equipment.

7 *ISDN*

Many engineers now feel that the increased use of LANs could well become one of the main factors leading to rapid integration of voice and non-voice services, providing us all with Integrated Services Digital Networks on an economic basis.

---

39 Integrated Services Digital Networks

An ISDN is a single network able to carry and switch a wide variety of telecommunications services. It is expected to evolve from an IDN, an integrated digital network, which is a telephony network in which digital transmission systems have been fully integrated with digital switching systems (fig. 39.1).

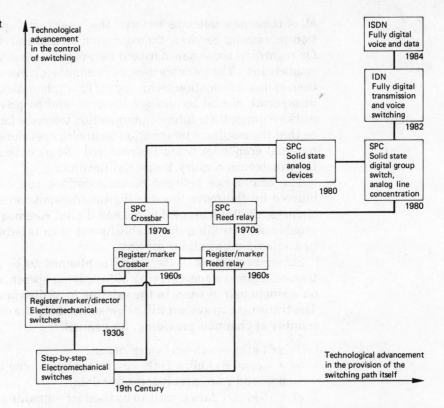

Fig. 39.1 The development of automatic telephony leading to the ISDN

Technological advancement in the control of switching

| ISDN Fully digital voice and data — 1984 |
| IDN Fully digital transmission and voice switching — 1982 |
| SPC Solid state digital group switch, analog line concentration — 1980 |

SPC Solid state analog devices — 1980

SPC Crossbar — 1970s

SPC Reed relay — 1970s

Register/marker Crossbar — 1960s

Register/marker Reed relay — 1960s

Register/marker/director Electromechanical switches — 1930s

Step-by-step Electromechanical switches — 19th Century

Technological advancement in the provision of the switching path itself

To most people who work in national telecommunications systems and have grown up in countries where telephone services are provided and maintained by a national monopoly, e.g. British Telecom or a PTT Department, it seems fair and reasonable to expect that this national service should be expanded to provide the public with all the new telecoms services which are now becoming available, such services as:

Slow-scan television.
Facsimile.
High-quality videotex (Picture Prestel).
Text transmission (teletex).
Computer–computer connections (circuit or packet switching).
Electronic funds transfer.
Telemetry (meter readings for power, water, etc.).
Interactive videotex (electronic shopping).
Electronic mail.

To other people this expectation is not always considered fair and reasonable at all. The computer industry has, for example, developed to its present strength as a largely privately-owned and competitive industry. Equipment and software made by one computer manufacture will, in general, not work with equipment from other manufacturers. There are very many private data networks in service all over the world; very few of these can be connected to each other or to the public switched network.

With the provision of complex national networks of computer-controlled telephone exchanges capable of switching, not only telephony but also all of these new telecoms services, the "traditional" providers of information processing services, the computer manufacturers, may be forgiven for regarding these new national networks as the beginnings of head-on competition. The convergence of communications and computing has been called information technology (IT) or telematics; as public networks incorporate digital technology features, and as private computing networks enlarge their intercommunication telecoms facilities, it could well be that the resultant information technology revolution will result in our national economies being restructured just as radically as they were in the nineteenth century Industrial Revolution.

The boundaries between communications and computing are being blurred by the move towards digital transmission systems, digital exchanges, digital storage devices and digital common channel signalling capable of controlling the establishment of calls between all the various new telecommunications services.

Subscriber access to the ISDN is planned to be provided by digital transmission systems carrying bidirectional speech, data and signalling on a single pair of wires in the existing local distribution cable network. The transmission system will allow subscribers the simultaneous use of a number of channels providing, for example:

a) a 64 kbit/s both-way voice circuit,
b) a second 64 kbit/s both-way circuit, available to be used either as a second voice circuit or for fast data,
c) a 16 kbit/s data circuit to be used for signalling and relatively slow data.

Connection and interworking of a wide range of terminal types will be possible, from a simple telephone instrument to a sophisticated computer or intelligent data terminal.

At the start of every ISDN call it is proposed that the terminal should provide information to the exchange indicating the procedures to be used to set up and control the call. Messages containing information more meaningful to data terminals or computers can replace the tones or verbal announcements which are acceptable to human users. If the service indication is "telephony", the call would not be restricted to the ISDN but could be switched through to anywhere in the worldwide telephone network. If the service indication is for a special service, it may be necessary to provide special interworking interfaces, e.g. to a packet switching network.

No one can yet say with any certainly how international ISDNs will develop but it seems fairly clear that the telecoms world is at the beginning of a period of extremely rapid development and change which will have a major effect on people's lives, everywhere.

Part H Digital Fundamentals

40 Bits, Bytes and Words

In section 32 it was shown how ordinary decimal numbers may be turned into binary codes so that they may be transmitted by a two-level pulse signal.

In computers we are concerned not only with figures but also with words; letters, punctuation marks, accents, etc., may all be given arbitrary values in an appropriate code so that, when that particular code is received, the printing device prints the appropriate letter or character.

The basic unit of information in a two-level digital notation is called a binary digit or bit. The contents of computer memories may be written or read in units, typically 16 or 32 bits long, each unit known as a word. 32 bits is long enough to represent a number to an accuracy of about ten decimal digits, enough for most calculations. A word may also represent one, two or more instructions, depending on the form of coding used. Another way in which a word may be interpreted is, as mentioned above, in terms of character symbols: thus a word of 32 bits might represent four characters, each occupying 8 bits, or a word of 24 bits might represent four 6-bit characters.

Normally a single reference to a memory will yield one word, though this is not an invariable rule. Every one-word location in the store has a unique number associated with it, known as its address, and it is by sending the address to the store that the control is able to "unlock" a particular location to gain access to it, i.e. to read out its contents, or to write a new word into that address.

A typical small memory block might contain 4096 memory locations. Notice the use of binary power here; 4096 is 2^{12}, so a binary number 12 bits long could be used to address any location within the block. 4096 is sometimes abbreviated to 4K, where the letter K denotes the binary kilo, i.e. 1024 or 2^{10}. The normal denary kilo of 1000 is, of course, denoted by k, as in km for kilometre. In the same way, M (which normally means a million or 10^6 but stands for the Greek prefix mega) is used to denote 2^{20} or 1 048 576. A typical memory for a mainframe computer has 20 binary million addresses (20 megabytes).

An 8-bit character is usually known as a byte. Some computers are byte-oriented, which means that a separate address is associated with every byte, or character. On small machines of this type, the store is actually arranged so that every reference produces one byte, though on larger machines a single store cycle will yield several bytes. Instructions and numeric data on byte-oriented machines consist of several bytes, the number depending on the accuracy required from the numeric item, or the amount of information the instruction has to supply.

> A BIT is the basic binary digit signal
> A BYTE is usually 8 bits
> A WORD is usually 2–4 bytes

Although bits, bytes and words are normally all that will be encountered, some microprocessor makers have recently designed equipment which addresses small units of only 4-bits instead of the more usual 8-bit byte. Somewhat coyly they have called this baby byte a nibble.

41 Integrated Circuits

An "integrated circuit" was originally the name given to several discrete components making up a distinct circuit stage, mounted on a separate printed circuit board. Then when it became possible to make up more than one circuit element on a single silicon chip (this was closely linked with the development of digital techniques and the increased use of transistors, logic gates, etc.), the name "integrated circuit" became applied solely to complete (and sometimes extremely complex) circuits all manufactured on a single base and packaged in a way which could readily be mounted on a printed circuit board.

The designers of these complex circuits aimed to reduce costs because ICs have a number of advantages over circuits using discrete components:

greatly reduced size
greatly reduced weight
lower costs
complex circuits become practicable

Unit costs do of course however depend greatly on the quantity of each type to be manufactured. In many cases the unit cost of a complex integrated circuit is now no greater than that of a single transistor, even though the one chip may incorporate hundreds of gates or other devices. There are no hard and fast dividing lines between different types of integrated circuit but a good guide to the number of gates that may be put down on a single chip and to the complexity of the various integrated circuits is:

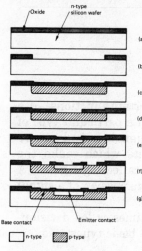

Fig. 41.1 The stages in the manufacture of a silicon planar transistor

(a)

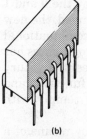

(b)

Fig. 41.2 (a) A planar transistor packaged as a discrete circuit; (b) The same transistor as a dual-in-line package

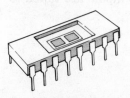

Fig. 41.3 Dual-in-line package with cap removed to show the cavity in which an IC may be mounted

SSI   small-scale integration

A single IC package containing from 10 to 20 gates, e.g. ICs containing say 4 AND gates and 6 inverters.

MSI   medium-scale integration

A single IC package containing from 20 to 100 logic gates or less than 100 memory bits, e.g. counters, shift registers, input/output ports.

LSI   large-scale integration

A single IC package containing from 100 to 5000 logic gates or 1000 to 16 000 memory bits, e.g. an 8-bit microprocessor chip.

VLSI   very-large-scale integration

A single IC package containing more than 5000 gates and possibly more than 100 000 memory bits, e.g. a 16-bit microprocessor chip.

Other abbreviations sometimes encountered in the integrated circuit world are DIL or DIP (dual-in-line package), PTH on PCB (plated through holes on printed circuit boards), and TCM (thermal control module: an IBM unit which accommodates up to 133 integrated circuits on a 28-layer ceramic substrate, having nearly 2000 pins per module and in effect water-cooled).

Although a great many transistors or other elements, together with their interconnecting links, are made on a single chip to turn it into an IC, the usual method of manufacture of monolithic ICs (mono = single, lithos = stone, in this case silicon) has many similarities in technique with the making of a single silicon (planar) transistor. The stages in one of the ways in which this can be done are shown in fig. 41.1.

a) A wafer of n-type silicon is oxidized to a depth of about 1 micron.
b) This oxide layer is then partially etched off.
c) The wafer is exposed to a vapour of the acceptor element boron. This element diffuses into the wafer to a predetermined depth (creating a p-type zone). The wafer surface is at the same time reoxidised.
d) A part of the newly reoxidized surface is then etched away again.
e) The wafer is then exposed to a vapour of the donor element phosphorous. This diffuses in (to create an n-type zone) and the surface is again reoxidized.
f) More etching away of the oxide surface follows, to separate what have become the base and emitter regions on the surface of the wafer.
g) Metal contacts are alloyed on to the etched areas.

The wafer is then cut to the required size, mounted on a suitable collector contact and leads connected to the base and emitter contacts. The finished transistor looks like fig. 41.2b.

An integrated circuit, as mounted on a printed circuit board, does not look very different from this single transistor although the package may be somewhat larger. With the cap removed (fig. 41.3), the small silicon chip will be seen inside, in its cavity; each of the dark rectangles may represent many thousand transistors.

Hand-held programmable pocket calculators, microprocessors, home computers, viewdata terminals—none of these would have become practicable propositions without the development of integrated circuits.

42 Logic Gates

Many electrical circuits are designed to perform certain functions—if a particular type of input is found to be present, then a particular type of output has to be transmitted. In our digital world, signals in and out are usually binary, i.e. 0 or 1. The electronic circuits which make the decisions and provide the appropriate outputs are called logic gates. A logic gate is therefore a circuit which allows the transmission of a signal when an appropriate input signal has been applied.

The logic gates we are concerned with are sometimes called decision-making or combinational logic gates; there are five basic types:

AND
OR
NOT
NAND
NOR

and two derived types

Exclusive-OR (or Anti-coincidence)
Exclusive-NOR (or Coincidence)

All of these gates have at least one input line and a single output line. The output can take only a 0 or 1 state; this is controlled by the 0 and 1 states present on the input lines. If any of the inputs is changed, the new output is completely determined by the new inputs. These combinational logic circuits have no memory. The design of circuits using logic gates has its own mathematics, known as Boolean, after the nineteenth century mathematician who developed a complete set of algebraic functions and rules to handle logic operations. Logic gates necessarily comply with these rules.

The operation of any logical element can be described by means of a Boolean truth table showing the output state of the circuit for all possible combinations of input states.

1 The AND gate can have two or more inputs. It gives a logic 1 output only when all its inputs are at logic 1. If any of the inputs are taken to logic 0, the output of the gate also becomes 0.

Truth table for a three-input AND gate

Inputs	A	0	1	0	0	1	1	0	1
	B	0	0	1	0	1	0	1	1
	C	0	0	0	1	0	1	1	1
Output	F	0	0	0	0	0	0	0	1

2 The OR gate can have two or more inputs. It gives a logic 1 output if there is logic 1 on any of its inputs. Its output is only logic 0 when all inputs are logic 0.

Truth table for a three-input OR gate

Inputs	A	0	1	0	0	1	1	0	1
	B	0	0	1	0	1	0	1	1
	C	0	0	0	1	0	1	1	1
Output	F	0	1	1	1	1	1	1	1

3 The NOT gate or invertor has only one input. Its output is always the logical inverse of the input state. Thus if the input is logic 0 the output is logic 1 and vice versa.

Truth table for a NOT gate

Input	0	1
Output	1	0

4 The NAND gate is a combination of an AND gate and a NOT gate, i.e. NOT AND, which is abbreviated to NAND. The output of a NAND gate is logic 0 only if all its inputs are held at logic 1.

Truth table for a three-input NAND gate

Inputs	A	0	1	0	0	1	1	0	1
	B	0	0	1	0	1	0	1	1
	C	0	0	0	1	0	1	1	1
Output	F	1	1	1	1	1	1	1	0

5 The NOR gate is a combination of an OR gate and a NOT gate, i.e. NOT OR, which is abbreviated to NOR. The output of a NOR gate is logic 0 if any one or more of its inputs is logic 1. Only when all inputs are held at logic 0 will the output go to logic 1.

Truth table for a three-input NOR gate

Inputs	A	0	1	0	0	1	1	0	1
	B	0	0	1	0	1	0	1	1
	C	0	0	0	1	0	1	1	1
Output	F	1	0	0	0	0	0	0	0

6 The Exclusive-OR gate (sometimes called the anti-coincidence gate) has only two inputs and gives a logic 1 output only when the two inputs are different. If both inputs are logic 1, or both are logic 0, the output is logic 0.

Truth table for an Exclusive-OR gate

Inputs	A	0	1	0	1
	B	0	0	1	1
Output	F	0	1	1	0

7 The Exclusive-NOR gate (sometimes called the coincidence gate) has only two inputs and gives a logic 1 output only when the two inputs are in the same logic state, either both 0 or both 1.

Truth table for an Exclusive-NOR gate

Inputs	A	0	1	0	1
	B	0	0	1	1
Output	F	1	0	0	1

There are several different types of integrated circuit logic gate, each type called a logic family because the members of the family share many design and operating features (and have also inherited many of their characteristics from earlier members of the same logic family). The logic families most commonly encountered in telecommunications applications are

TTL   transistor-transistor logic
ECL   emitter-coupled transistor
CMOS complementary metal oxide semiconductor

a) TTL gates depend on the interaction of several transistors; the input is usually fed through one transistor and the output obtained via a pair of transistors in push-pull operating mode.
b) ECL gates depend on the operation of a transistor pair coupled together by their emitters.
c) CMOS gates use metal oxide semiconductor field effect transistors (MOSFETS) as electronic switches. Because load resistances of high values are needed for MOSFETS, and it is comparatively expensive to fabricate these in an integrated circuit, these high load resistances are provided by using a second or complementary MOSFET.

In both the TTL and CMOS families, NAND and NOR gates are the most commonly used since their cost is less than that of the other gates. Also, NAND and NOR gates generally have a faster operating speed and lower power dissipation. Very often therefore, a digital circuit is composed only of NAND and/or NOR gates.

Simple circuits using discrete components (mostly resistors and transistors) could be used to make up any of the logic gates in any of the family configurations but they would be totally impracticable for most digital systems where hundreds of gates are needed. Integrated circuit versions are therefore commonly used.

The properties and parameters usually used in specifying gates are:

FAN-IN—the number of inputs that can be accommodated on one gate.

FAN-OUT—the ability of a gate to drive several inputs of similar gates simultaneously.

NOISE MARGIN—the measure (in volts) of a noise signal that can be accepted without causing a change of state at the output.

POWER CONSUMPTION—the amount of power taken by one gate during static (i.e. steady state) and dynamic (switched) conditions. The greater the power dissipation of a gate the more heat must be removed from the circuit.

PROPAGATION DELAY (or SPEED OF OPERATION)—the time that elapses between the application of a signal to an input terminal and the resulting change in logical state at the output terminal.

The TTL logic family is at present the most widely used for all logic gates, primarily because it offers fairly high speed, good fan-in and fan-out, and is relatively cheap and available from many sources. It does however have poor noise immunity and a rather high power consumption. A low power consumption version is available but the speed of operation of this variant is poor. Schottky TTL logic has however now been developed; this involves the connection of an internally-provided diode between base and collector of the transistor which makes for significant improvements to basic TTL parameters, providing fast operation, low power dissipation, and high-frequency capability. The latest developments in this family are called Advanced Low-power Schottky.

The CMOS logic family offers very low power dissipation and good noise immunity but at present has the disadvantage of relatively poor speed of operation.

The ECL logic family offers extremely fast speed of operation and finds favour when maximum possible speed is the prime consideration. Some of the latest Advanced Low-power Schottky TTL units are however now able to offer comparable speeds.

The table below summarizes the differences between these logic families.

	Propagation delay (ns)	Power dissipation (mW)	Noise immunity (V)	Fan-out	Fan-in	Supply voltage (V)
TTL	9	40	0.4	10	8	5
TTL Schottky	3	40	0.3	10	8	5
TTL Low-power Schottky	8	8	0.3	10	8	5
CMOS	30	0.001	1.5	50	8	5
ECL	1	30	0.4	50	5	−5.2

43 Memories and Stores

In most digital systems and in all computer-type applications, elements capable of storing binary-coded information are required. For example, the arithmetic/logic circuits of a computer must be able to hold results during the processing of a problem, and memories in a computer-controlled telephone exchange must hold all available information about each line and the equipment connected to it.

It is therefore of fundamental importance to provide logic elements which have a memory and are capable of storing binary data for a given time so that this may be available for subsequent use when needed.

A computer usually requires different kinds of memory stores:

a) A main store: this contains all the information which must be immediately available to the processor.

b) A larger capacity backing or back-up store: this contains most of the data and programmes and is used to hold information that need not be immediately available to the processors.

c) Several registers: these provide short-term storage of information such as interim results of individual stages in complex calculations, or data read out from a main or backing store so as to be ready to be used by a processor.

It must be possible both to write information into a memory and to read information out of the memory. To make this possible a memory consists of a large number of locations—often arranged in a matrix—in each of which a small amount of data can be stored. Each location has a unique address so that it can be accessed from outside the memory. The access time of a memory is the time that is needed to read one word out of the memory, or to write one word into the memory.

The main stores in early computers were usually made up of a matrix of ferrite cores; these are small rings of ferrite through which wires are threaded which both control the magnetization of each core and read out the direction of such magnetization. Main stores very rarely use ferrite cores nowadays because semiconductor stores have become cheaper and are far more compact—but the name has remained in use, so a main store is often called a core store even though it uses no cores.

Magnetic tape and magnetic disc are widely used, especially for back-up stores where rapid and equal time access to all the locations in the store is not so important as the ability to hold information irrespective of the failure of power supplies. Both these magnetic stores work on the same principles as the domestic tape recorder and are described in more detail in section 44.

The magnetic bubble is beginning to come into use in special applications such as telephone exchanges where an extremely large storage capacity, rapid access, and no moving parts to require maintenance are of paramount importance. Bubble memories are at present expensive devices; their manufacture demands extremely high precision.

Solid state semiconductor memories are however the types of memory which are of most immediate concern to us, they are very widely used in electronic equipment of all types.

Basic semiconductor memory elements are bistable, and are often called flip-flops or latches. When used to store a logic 1 bit, the input "flips" the memory element so that the output takes and holds the logic 1 state. It will hold this state indefinitely or until it is reset or "flopped" back to its other stable state, the logic 0 state, by a subsequent input. The several varieties of flip-flop are made in integrated circuit form, and many IC chips contain a large number of circuits made up of combinational or logic gates interconnected with memory elements.

Most systems can operate with both volatile memory (memory which forgets or loses its contents when the power supply is removed) and non-volatile memory (which retains its contents even when power is removed). In general, volatile memory is used as workspace for storing temporary data and the results of calculations during program execution, whilst non-volatile memory is used to store programs, permanent records, etc.

There are two basic types of semiconductor memory device:

1) Those into which information has been written during manufacture; this information can be read out by the user but it cannot be changed. These are called Read-only Memories or ROMs.

2) Those into which information can be written by the user, changed by the user if necessary, and read out by the user when wanted. These are called Random-access Memories or RAMs. This is not really a very good descriptive name for these devices; it is perhaps possible that, if their original designers had called them Write And Read memories (which they are), they might, as WAR Memories, never have become socially or economically acceptable in some quarters.

A ROM, then, is a memory which is permanently programmed and can only be read, i.e. program instructions, data, etc., can only be copied from ROM memory because each ROM has its information "burnt" into its store during manufacture. Typical applications are

a) Monitor programs or operating systems programs, which are the programs that control the operation of a microcomputer system and allow the user to run application programs, to input and output data, to examine and modify memory, etc.

b) Dedicated programs, each one specially devised to fulfil one particular function, such as a control program or a complex calculation routine.

ROMs are usually programmed in the factory and, since it is clearly not possible to maintain power supplies to them while they are being transported for assembly into a system, they have to be non-volatile; information stored in a ROM is not lost when power supplies are switched off.

A RAM is designed to have information both written in to it and read out of it. Random access means that each stored word can be accessed or retrieved in a given amount of time. The memory locations storing the words making up a program instruction, data, etc., can be accessed in random order in the same amount of time, independent of the position of the word in the store. In this sense, ROMs are also random access devices but programs stored on magnetic tape must be accessed sequentially and are therefore not random access. RAMs are used for storing application programs during use, fed in for example from a tape, disc or printer. They are also used for storing intermediate results during a program execution. Information stored in RAM can readily be modified. RAM memory is usually volatile and lost when power supplies are switched off; if a back-up power supply maintains power in the event of mains failure, the information in store will of course not be lost.

The ROM has been described in its basic form. There are, however, a number of important versions of it, three of which are now briefly described.

A programmable read only memory or PROM is a special type of ROM designed to be programmed by the user to suit a specific application for a device. It uses a column-and-row memory matrix with all of the inter-sections linked by fusible diodes (transistors). It is programmed by addressing the particular locations in memory that are to store logic 1 and passing a sufficiently large current through the transistor at that intersection to blow the fuse, so that there is no longer a link between column and row of the matrix at that particular point.

An Erasable PROM or EPROM is similar to an ordinary PROM except that the chosen matrix intersections are not permanently made logic state 1 by blowing internal fuses. This logic state is written in by the storage of an electrical charge at the point. If a program on an EPROM is to be changed, the EPROM is exposed to ultraviolet radiation directed through a small window in the package containing the chip. This removes the stored charge at every location in memory so the chip has to be completely reprogrammed.

In an Electrically-alterable PROM or EAPROM, as with the EPROM, programming a memory cell to store logic state 1 is accomplished by charging that cell. With an EAPROM, erasure procedure is carried out by applying a reverse polarity voltage to a cell to remove its charge; this erasure is done on individual cells in the matrix, it is not a complete chip erasure as it is with an EPROM.

44 Magnetic Tapes and Discs

Tapes and discs use the same physical effect, that of magnetic surface recording, to store information. The only essential differences between the systems are the mechanical arrangements of the read/write circuitry and the recording medium.

1 *The Principle* (fig. 44.1)

The "write head" consists of a coil wound on a core material which has a high magnetic permeability. The air gap in the core allows magnetic flux, produced by the write current, to pass into the magnetic surface of the recording material, resulting in a small area of the surface becoming magnetized. Reversing the write current results in a reversal of the magnetization, thus allowing a binary pattern to be recorded.

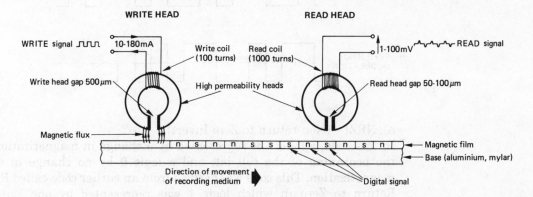

Fig. 44.1 Principle of magnetic surface recording

The "read head" is constructed in a similar fashion to the write head. As the magnetized surface passes the read head, some of the magnetic flux passes into the head via the read-head gap. The changing flux induces a voltage into the read head windings which reflects the state of magnetization (and hence the data recorded) of the recording surface.

The only essential differences between the two heads are in the number of turns on the coils and the sizes of the air gaps. The write coil has a small number of turns (giving low inductance and high writing speed); the write head has a large air gap (to minimize write current). The read coil has a large number of turns (to produce adequate signal level); the read head has a small air gap (to provide good resolution).

The recording surface is a thin film of ferric oxide, nickel-cobalt, or chromium dioxide. Of these the iron oxide is cheaper but not quite as effective.

2 *Magnetization Codes and Error Detection* (fig. 44.2)

There are several different magnetization codes in use; the most common are the following.

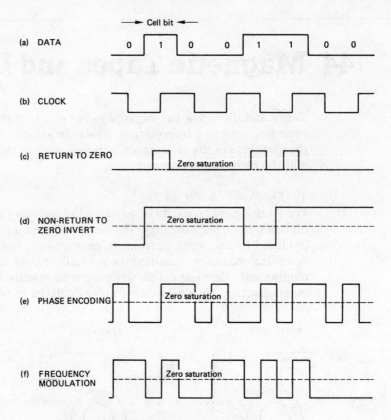

Fig. 44.2 Magnetic surface recording codes

(a) DATA

(b) CLOCK

(c) RETURN TO ZERO — Zero saturation

(d) NON-RETURN TO ZERO INVERT — Zero saturation

(e) PHASE ENCODING — Zero saturation

(f) FREQUENCY MODULATION — Zero saturation

Cell bit

0 1 0 0 1 1 0 0

a) NRZI (Non-return to Zero Invert)

In this code, a logic 1 is represented by a change in magnetization at the beginning of the cell bit, and a logic 0 by no change in such magnetization. This code was derived from an earlier code called RZ or Return to Zero in which logic 1 was represented by one state of magnetization for the central region only of the cell bit, with magnetization reducing to zero for the rest of the cell bit. Logic 0 was represented by no magnetization.

b) PE (Phase Encoding)

In this code, logic 1 is represented by a change of magnetization in a specified direction at the centre of the cell; logic 0 is represented by a change of magnetization in the opposite direction. PE has the great advantage that regular signals are given at the centre of each cell and so can be used for synchronization without needing an external clock. If NRZI is used, however, a separate synchronizing or clock signal has to be provided in order to keep the receiving device exactly in step with the coded information on the tape.

c) FM (Frequency Modulation)

In this code, logic 1 is represented by a change of magnetization at the centre of the cell and logic 0 by no change. In addition a change of magnetization occurs at the beginning of each cell. These changes mean that the system, like PE, is self-clocking.

Magnetic surface recording is susceptible to the generation of errors

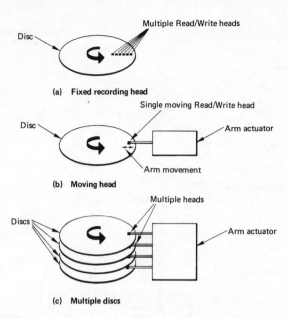

Fig. 44.3 Rigid (hard) disc systems

(a) **Fixed recording head**

Disc

Multiple Read/Write heads

(b) **Moving head**

Disc

Single moving Read/Write head

Arm actuator

Arm movement

(c) **Multiple discs**

Discs

Multiple heads

Arm actuator

and it is usual to incorporate in the read/write process some form of error detection. The most common methods are:

a) Parity checks. A parity bit is added to each word during the write operation such that the total number of 1s in each word is either even or odd. This is checked during the read operation.

b) Check sums. The numerical value of each byte in a block of data is summed and the total written at the end of the data block. During reading, the check sum is again calculated and compared with the total written in.

c) Cyclic redundancy check. Each block of data is treated as if it were one long number. This number is divided by a known and agreed number called the generator. The remainder is recorded at the end of the data block. During reading, the calculation is carried out again and the remainder obtained compared with that written in.

3 *Magnetic Discs and Tapes*

The recording surface in a disc system comprises a disc which revolves around an axis, usually vertical. Both heads are close to the disc surface. Disc access times clearly depend on the speed of revolution of the disc. Heads do not rub physically against the revolving disc but are separated by a very small air gap. Freedom from dust is therefore extremely important.

Rigid or hard discs (fig. 44.3) are usually associated only with large computers, and are made of aluminium coated with the recording surface, often chromium oxide. These discs are available with a storage capacity per disc of up to 2 Mbytes. Typical discs have up to 256 separate tracks and in some models there is one fixed head (combined read/write) for each track, i.e. 256 separate heads over each disc. These give very small access times. Other models have moving heads which have to be precision

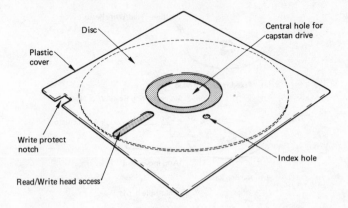

Fig. 44.4 A floppy disc

Disc

Central hole for
capstan drive

Plastic
cover

Write protect
notch

Index hole

Read/Write head access

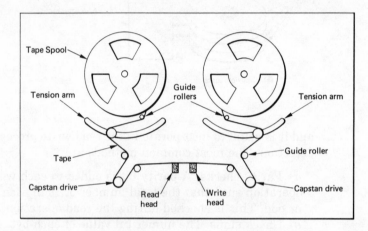

Fig. 44.5 Magnetic tape
transport

Tape Spool

Tension arm

Guide
rollers

Tension arm

Tape

Guide roller

Capstan drive

Read
head

Write
head

Capstan drive

engineered in order to be driven to exactly the right track. Some moving head units are made with up to 12 stacked revolving discs, each with its own head unit—these allow action to be taken on a parallel basis, so give much faster access times than single disc moving head units. Some of these multiple disc units use double-sided recording and have a total capacity of 30 to 40 Mbytes per system.

Floppy discs (fig. 44.4) are small flexible discs about the size of a 45 rpm record with the recording surface of ferric oxide on one or both sides. Floppy discs are protected by plastic sleeves, with a radial slot which gives access to the read/write heads. Some floppy units have heads which ride in actual contact with the discs themselves; this limits the life of a floppy. Standard floppies have a capacity of 4 Mbytes; specials are however available with double this capacity.

Tapes usually have a polyester base with a thin layer of ferric oxide as the recording surface. Different tape widths and reel diameters are used for different purposes. There are usually nine tracks across the width of a tape. The tape normally maintains contact with the multi-track heads as it is driven past them (normally at about 1 metre per second) so there is some tape wear due to friction. The main cause of limited life for tapes is however the mechanical fatigue caused by continual acceleration and deceleration of the tape during searches. This points to the main drawback of tape—it is a serial-access medium, whereas a disc is random access. A simplified drawing of a typical tape transport system is given in fig. 44.5. Tape reels usually hold 500 metres of tape, giving typical maximum storage capacity of 100 Mbytes.

45 Computers and Microprocessors

Digital computers are basically electronic calculating and data processing machines which work with instructions and data coded in simple binary digit form. Although computers were originally purely mathematical devices, the use of logic enables many non-arithmetic tasks to be carried out under computer control. Computers have a wide and ever-increasing number of applications.

Digital computers, shown in schematic form in fig. 45.1 may be loosely classified into three broad classes, based on cost and degree of sophistication:

a) Microcomputers—from £40 ($50) upwards
b) Minicomputers—from £1000 ($1500) upwards
c) Mainframe computers—from £100 000 ($150 000) upwards.

A microprocessor is an integrated circuit containing all necessary computational and control circuitry on a single silicon chip, and is the brains of a microcomputer. Microprocessors control washing machines, petrol pumps, word processors and electronic games and are often used in telephone exchanges to perform tasks which were at one time carried out only by hard-wired circuits. For example, micros scan groups of subscribers lines so that as soon as a subscriber goes "off hook" the necessary receiving equipment may be connected to the line to receive the dialled-in digits. A typical 10 000 line digital telephone exchange of modern design is likely to utilize almost 1000 microprocessors, all operating on a distributed control basis. Most microprocessors operate with an 8-bit (1 byte) data unit, each unit representing a character signal.

A minicomputer is used for more extensive applications where, for example, greater speed, greater memory storage or more diverse control functions are needed. Most minis operate with a word length of 12–16 bits, the equivalent of two characters.

Mainframe computers are large-capacity machines used for large-scale commercial data processing or for running simultaneously a great many remote computing terminals. Mainframe computers operate with a word

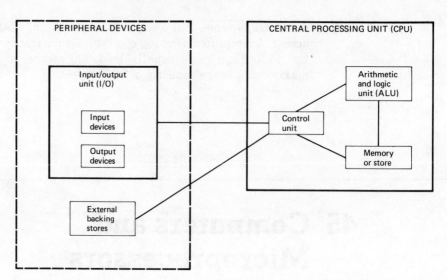

Fig. 45.1 Basic schematic of a digital computer

PERIPHERAL DEVICES

Input/output unit (I/O)

Input devices

Output devices

External backing stores

CENTRAL PROCESSING UNIT (CPU)

Arithmetic and logic unit (ALU)

Control unit

Memory or store

length of 20 or more bits per word; they are able to carry out complex mathematical calculations without errors building up.

Regardless of size most digital computers are made up of

a) Central processing unit (CPU)
b) Input and output devices
c) Auxiliary units.

a) *Central Processing Unit* Although the three main functions are shown in fig. 45.1 as separate boxes, all three functions in some microprocessors are carried out by a single integrated circuit chip.

The Control Unit interprets and carries out the instructions of the program as held in the memory unit. The control unit sometimes incorporates the clock, which is usually a very stable crystal-controlled oscillator which feeds pulses to all circuits in the system. Sometimes the clock is a separate subsystem in the CPU.

The Arithmetic/Logic Unit (ALU) performs mathematical calculations and makes logic decisions on the data fed to it from the memory unit. The most important of the logic decisions are the three comparison operations of greater than, less than, and equal to. It is the ability of a computer to perform these operations which contributes substantially to its usefulness and power.

The Store or Memory Unit holds the data and instructions of the program which have been fed into the CPU via the I/O unit. It also holds the permanently stored programs (called the Operating System) that enable the computer to function.

b) *Input and Output Devices* enable the user to communicate with the computer and thus provide the man/machine interface.

Input devices include:
  Keyboards
  Tape cassettes
  Disc drive units

Output devices include:
  Printers
  Visual display units (VDUs)
  Audible tones
  Lamps

Fig. 45.2 Simplified architecture of a microcomputer using a microprocessor

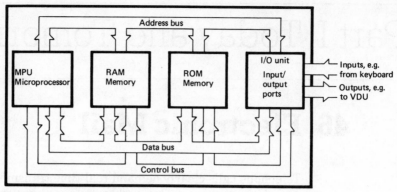

(a) A basic microcomputer system

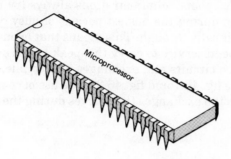

(b) A typical microprocessor
Integrated circuit dual in line (D I L) package, with 40 pins for external connection to data, control and address buses.
(Drawn to scale: approx 52 x 13 x 3mm)

Fig. 45.3 Units making up a microcomputer system plus some typical peripherals

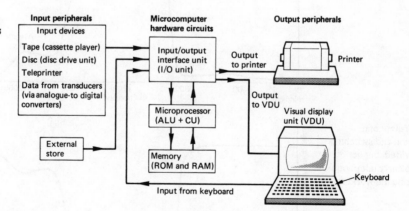

Part I Today and Tomorrow

46 Electronic Mail

Telephone calls clearly demand real-time transmission; on the phone we expect our listener to hear what we say as we speak and to be able to reply straightaway.

Telephone administrations always try to plan their networks so that, even during the busiest periods of busy days, callers will be able to get their calls through. This means that enough circuits and exchanges have to be in service to carry this peak traffic even though for much of the day these circuits and exchanges will be idle, not really earning their keep. (See fig. 46.1 and fig. 46.2.) The use of some of these circuits and possibly also the exchanges themselves during the idle periods will therefore help

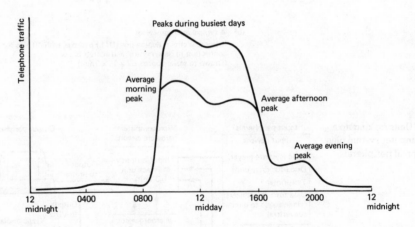

Fig. 46.1 Variation of telephone traffic during the working day

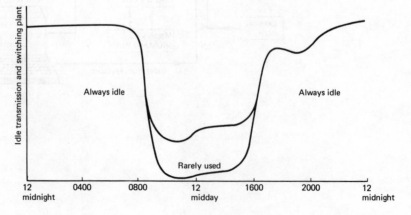

Fig. 46.2 Telephone transmission and switching plant provided, but not needed to carry telephone traffic during the working day

the administration—the facilities have to be there, so why not use them and earn extra revenue.

The cost of collecting, sorting and delivering ordinary mail continues to rise; these processes are necessarily somewhat labour intensive in most countries. Such mail does not have to be delivered within minutes of its despatch; we are usually more than happy if letters are delivered during the day following their being posted. It has been calculated in the USA that only about 30% of their first class mail is personal in nature; about 70% represents business correspondence such as accounts, orders, receipts, invoices, payments, etc. Almost all these are prepared by a computer in the company of origin and are dealt with by a computer in the company to which they have been delivered.

Now that many national administrations are changing their transmission networks over from analogue (FDM) to digital (TDM) working, it is comparatively straightforward to arrange for one computer to work direct to another. Why go through the stage of printing a bill, to be sent physically through the mail, when the information could easily have been passed from one company to the other, without being manually handled in this way? In some organizations it may not in fact be necessary for a paper copy of a bill to be prepared at all; any necessary checking and passing for payment could be done using a visual display unit and a remote terminal on the company's computer.

If the telephone network is to be used for the transmission of material which does not necessarily have to be dealt with on a real-time basis, there will have to be an indication given of the nature of the offered traffic to ensure that telephone calls themselves are put straight through. These other lower-priority tasks can then be dealt with in bursts of information whenever the circuits would otherwise be idle. The type of signalling facility needed to control such services is already available with CCITT's common channel signalling system no. 7. This can be used to control the setting-up and supervision of all types of calls, both voice and non-voice.

It seems likely that for a good many years yet, very few households will find it economic to equip themselves with special apparatus to receive electronic mail direct. Such mail is likely to be printed out on high-speed printers in the local Post Office and delivered by messenger.

It is a salutary thought that a human voice signal carried on a PCM system is represented by 64 000 bits per second. If a telephone call lasts three minutes, almost 12 million information bits will have been sent in each direction during the call. A typical bill being sent by electronic mail is, for comparison, likely to need only three or four thousand bits, and these could be transmitted whenever the channel is idle. It will be very easy to use national telephone networks to deliver such mail electronically, far more speedily than it can at present be delivered.

Not very many years ago, all financial transactions between banks were carried out by mail. Then when public telegraph systems were introduced, they were used. Nowadays, many major financial transactions are carried out using the switched data network of the Society for Worldwide Interbank Financial Telecommunication, SWIFT (see fig. 46.3). In one way, this could be considered to be indicative of the type of service which electronic mail could provide for many other industries—SWIFT aims at providing a secure, standardized, auditable, controllable, and rapid method for member banks to effect financial transactions. Because messages often deal with huge sums of money, security and control receive special attention, more than would normally be appropri-

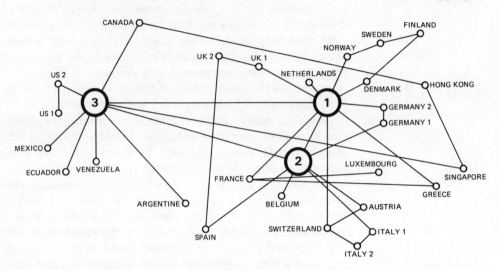

Fig. 46.3 Data networks for the Society for Worldwide Interbank Financial Transactions (SWIFT). The three principal operating centres are in the Netherlands (1), Belgium (2) and the USA (3)

ate for an electronic mail system. For example, if an urgent message has not been delivered to its addressee anywhere in the world within 20 minutes, the sender receives an "overdue" warning advice explaining the reason for non-delivery.

Many companies now send advertising circulars by mail to large numbers of addressees. Such bulk distribution could be carried out very economically if copies of the circular were printed out by electronic mail at each Post Office responsible for delivering the circular to clients. Another comparable type of service which is being planned in many areas is a one-way voice message. The message to be delivered is recorded and sent forward to a store in the exchange together with a list of telephone numbers to which the message is to be given (Section 47).

More and more items which are now being moved physically by mail will, during the next few years, be transmitted electronically, with final delivery to destination by a medium appropriate to the message—on paper, by word of (automatic) mouth, or on to a TV screen.

47 Voice Messaging

Most of the famous messages in our history books are one-way signals to which no effective answer was possible—the simple message "NUTS" issued by a US General when the Germans invited him to surrender at Bastogne in 1944 is a classic example of this.

The majority of telephone calls (apart from purely social gossip calls) are originated because people either want to tell someone something, or to ask someone to do something. There may of course be lots of interactive chit-chat mixed up with basic requests, questions and answers but the main reason for each call is usually fairly specific.

Telephone people all over the world are perhaps a bit like confidence tricksters: we have persuaded others that they really do need both-way conversation, when all most people basically want (in the business world at any rate) is to be able to deliver a message which will be acted upon. If the distant-end person is difficult to contact, it could be that, by the time you get through, the person has just gone out to a meeting. The return of your call finds that you are with the Chairman and cannot be disturbed. Your return call just fails to stall a prompt departure to Miami for the weekend—all good business for the telephone company but not really very productive for you!

The ordinary telephone answering machine was an early way of dealing with this difficulty; remember that the tape recorder is a relatively new invention, it was not until the 1960s and 70s that these became widely available. An answering machine is connected to your personal telephone so that, if you do not pick it up after the third ring or so, it starts up and informs the caller, usually in your own voice, that you are away until tomorrow, please call back later. The next refinement was to equip the answering machine with the ability to record: "please say who you are and leave your message as soon as you hear the 'beep' tone". When you get back to your office, you play the whole tape back (or your secretary types out a list of callers' names and messages) so that appropriate action can be taken. With more sophisticated machines you can call in from outside and command your machine to play back the received messages to you, over the phone.

Strangely enough, overall public reaction to answering machines is not usually particularly enthusiastic. People who have them say they are useful but a great many callers somehow seem shy about talking to a recording machine; they just mumble "sorry", and hang up.

But machines like these are here to stay so we'd best all try to get used to them. A telephone-based sales campaign, for example, uses the same message hundreds or even thousands of times. It would be soul-destroying to expect a human being to give out identical messages to a thousand listeners, ignoring the fact that these sales messages often generate rude attempted interruption by the captive listener. Machines become especially useful if there are time zone differences: a businessman in Singapore has a sudden thought to pass to a colleague in Chicago, but it is 17.00 in Singapore and only 03.00 in Chicago so an ordinary telephone call would not be very popular, unless of course big money was involved. Much better to phone this message through to a machine in Chicago which will pass it on at a civilized hour later in the morning.

A user of a voice messaging system (VMS) is pre-geared to talk to a machine. The psychological hurdle (associated with ordinary answering machines) of expecting a human reply but getting a click and a non-interruptible robot does not have to be jumped. VMS users can certainly expect lower long-distance call bills than those who insist on person-to-person contact. That basically is what todays voice messaging systems are all about. VMS rationalizes the concept of the telephone answering machine and makes these facilities available either to all company officers (for in-house messaging systems) or to anyone who joins a particular commercial Voice Messaging Service. You dial the number and repeat your message. The system tells the addressee that there is a message waiting in the "mailbox", and by keying in a command digit, the message is played back and then erased.

Technical and procedural aspects of voice messaging are at present (1985) changing extremely rapidly. Manufacturers are continually developing newer and more complex devices. There are no international standards yet but almost all systems depend on the controlling phone being able to send push-button-dialling voice-frequency tones, so if you only have a rotary dial service it may be difficult for you to find an economic voice messaging system to suit your requirements.

One of the reasons for growth of voice messaging not being as explosive as its proponents might wish is a purely social one: telephone calls, say to a branch office, are very rarely completely impersonal, there is usually some small element of chat even in the most serious and official of calls. In a military environment it is no doubt essential that areas of responsibility should be so closely defined that a brief command "down the ladder" and an acknowledgement or report back up again are sufficient, but in a normal civilian world there are usually very many lateral relationships to be considered as well as purely hierarchic ones. Very few major decisions can be made in an isolated way. It is therefore worth mentioning that voice messaging does not really mean that the social side of business calls has to be cut right out but the realization that you are talking to a machine and not personally to your friend undoubtedly does have something of an inhibiting effect in some circumstances.

Most present-day message systems use magnetic tape recording but technology now permits messages of a minute or so to be stored economically in binary form on chips, i.e. with no moving parts. It might be interesting at this point to note the present position in the closely allied fields of voice recognition and artificial voices, sometimes confused with voice messaging.

Several different systems have been designed to produce human speech artificially, without having to record each phrase separately and splice words together to make up sentences. The UK's System X digital switching system includes a compact computer-controlled voice guidance unit which can build up a large number of sentences from a chip store of digitally encoded phonemes. For example, customers using Speed Dialling (Abbreviated Dialling) are advised step by step in a friendly and helpful way how to go about setting up their calls. Some modern cars have single-chip artificial speech units which remind drivers to "Fasten your seat belt" and give other warning or advice phrases at appropriate moments, and some aircraft have generally similar warning messages such as "Retract undercarriage", reinforcing normal instrumentation. Many of these artificial speech devices treat voiced sound (such as vowels) differently from sounds which depend on tongue and lip movements (like explosive B sounds), and many such devices produce voices which are perfectly acceptable and difficult to tell from a genuine human voice.

Much research is being carried out in attempts to transmit speech using the lowest possible bit rate. 64 kbit/s as used in CCITT PCM produces extremely good quality speech; in some military equipment, bit rates as low as 8 kbit/s have been used for speech, possibly sacrificing voice quality for security.

Voice recognition is also now receiving a tremendous amount of research attention. Several laboratories have produced units which will recognize a restricted number of words when clearly spoken and can if necessary immediately type out translations into other languages, or send out a digital message which can be read on a distant VDU screen as the words concerned. Voice recognition equipment is at present, however, not readily available in the market place; it is likely to be many years before such products become common.

48 Teletext, Viewdata, Prestel

Teletext is the name given to the system which permits a limited number of pages of text to be transmitted by TV broadcasting stations together with their program emissions; these special signals are transmitted using lines which are not being used for the ordinary video signal. A special decoding unit in domestic TV receivers permits selection and display on the screen. All the information is transmitted on a cyclic repetition basis by the broadcasting station without any need for communication back to the TV transmitter from the domestic receiver. There is bound to be some delay between choosing a particular information frame and the display of that frame on the screen; to minimize this delay a restricted number of frames of information are transmitted, only a few hundred.

Teletext services are provided in Britain by both the BBC and ITA: Ceefax and Oracle. In addition to the separate information frames, some broadcast TV programmes are accompanied by a teletext display of spoken dialogue so that these TV programmes may be enjoyed by the hard-of-hearing.

Viewdata, sometimes called videotex, is somewhat similar in appearance on the screen but is an interactive service. Viewdata has been operational in Britain since 1979. The UK service is now called Prestel; it uses a slightly modified domestic TV receiver in conjunction with a completely normal public telephone line in order to provide an interactive computerized data retrieval service (fig. 48.1). Generally similar systems have also been designed in France and in Canada.

Prestel service is provided in Britain by a network of computers organized to store frames of information which are transmitted (over the telephone line) at 1200 bit/sec when requested by the customer. A reverse channel at 75 bit/sec is provided to allow the customer to select the frame required or to provide answers to questions set in the information frame itself. This bit rate is sufficient for manual keyboard operation.

The information on the database is provided by a large number of specialist companies. Some pages of information have to be paid for, some are free. Some companies put information on the database on a restricted access basis, e.g. only Prestel customers from their own organization are permitted to call up these particular frames.

One major difficulty at present is that the viewdata systems in use in different countries use different procedures, codes and signals so they cannot all interwork with each other: customers using British Prestel cannot, for example, access the Canadian Telidon system. The Prestel system was the first of its type to be developed and uses what is called an alpha-mosaic method of constructing letters from small squares of colour; that is why they seem a bit odd to those who are used to normal TV definition pictures. Telidon, much later in the field, uses an alpha-geometric method of constructing letters, with lines and curves as well as squares of colour, so it is able to give an easier-to-read page of text and better diagrams than Prestel. Some countries are now experimenting with hybrid viewdata/teletext systems by which the interactive or "command" signals from the customer to the computer are sent in on a telephone line but the called-for page of information, either text or picture, is transmitted by a radio signal, coded, stored and displayed only by the customer asking for this particular page.

Considerable activity is taking place in a number of international bodies to define standards which would permit interworking between generally similar services around the world. Among the bodies involved in these discussions are the CCITT (International Telegraph and Telephone Consultative Committee), the CCIR (International Radio Consultative Committee), the ISO (International Standards Organization), the CEPT (Conference of European Posts and Telecommunications Administrations), and the EBU (European Broadcasting Union).

Improvements are continually being made to Prestel services; one which will have a significant effect is Picture Prestel which allows the insertion of a still picture of colour television quality into an information page. The only restrictions which affect this are the cost of the extra memory required in each terminal, and the time taken to transmit the required amount of data on a low-bandwidth telephone circuit. When digital transmission systems and digital exchanges such as System X have been brought into wide-scale use it will be possible for high-

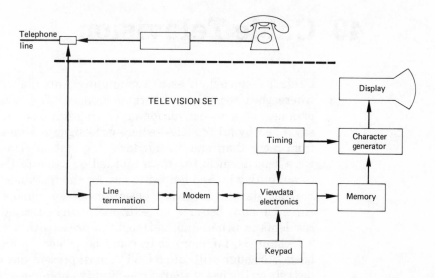

Fig. 48.1 A typical Prestel terminal

definition Picture Prestel to be transmitted very much more rapidly than is now possible.

Prestel Gateway is an extension of ordinary viewdata services which provides for greater interaction with users and enables Prestel customers to obtain access to computer networks managed by other organizations. For example, among the banking services available to Prestel Gateway users are

Accessing statements and ordering chequebooks.
Handling loan enquiries and showing repayment options.
Stopping payment on cheques.
Amending standing orders.
Effecting transfers of funds.

Airline and tour operators are among those who find Prestel interactive services most useful; it is possible now to display on the screen all possible alternative flights and for tickets on a particular flight and for accommodation to be booked, all by a single call.

Prestel users are able to do their shopping via supermarket computers, examine the choice of available items, check prices, make a note of special offers and then order the goods, all without leaving the comfort of their own homes.

49 Cable Television

Cable TV started off as a "Community Antenna" TV service. In towns where good reception of normal broadcast TV was not possible, local arrangements were made for a receiving station to be built at a location—say a nearby hilltop site—where good signals were available, preferably from more than one TV broadcasting station. The signals so received were sent down to the town and fed by cable to those residents in the community who had helped to pay for the installation.

What started as a very basic service for remote rural areas has in America now become an extremely sophisticated service, providing residents in urban and metropolitan areas with access to a great many TV channels, far more than could be picked up on a purely broadcast basis. Although still called CATV, most present-day cable TV services do not rely on the use of shared community antennae at all. Programmes are fed directly from studios or TV switching centres into the cable networks.

In the USA there is one particularly important feature which has encouraged the growth of cable TV: America has so many tall steelframe buildings that the reception of broadcast TV in major cities is rarely entirely satisfactory. Multiple reflections from skyscrapers between the TV station and the domestic receiver often mean that pictures are full of ghosts and peculiar colour combinations. About half the homes in the USA now watch cable TV rather than direct broadcast TV. Many of these systems are able to provide twenty or more different TV programmes.

Very few ordinary households need to be able to choose between so many possibilities, although this ability to carry many channels does make it practicable for programmes covering purely local interests and minority pursuits to be distributed. It is however in other possible uses of cable systems that there is the greatest potential.

On the cultural side, a hundred high-quality music channels could be carried in the bandwidth needed for a single colour TV channel; it would be relatively easy to satisfy classical music lovers. Many hundreds of voice-only programmes could similarly be distributed, enabling instructional programmes on all conceivable subjects to be made available.

Cable TV systems installed in the USA since 1972 have nearly all been provided with reverse-channel capability; this means that the customers can respond or reply to questions asked or can choose particular programmes to be transmitted. There is often also provision for some special programmes such as first-run movies to be available, these only to customers who pay a special fee for them. Among services now being planned are the transmission of newspapers by TV using a special method of transmission and a local memory store which will enable a single page of newsprint (or a close-up of columns) to be displayed on the screen for as long as the customer needs it.

In Britain the Prestel service has provided interactive data retrieval facilities for some years but this service has not yet become widely used,

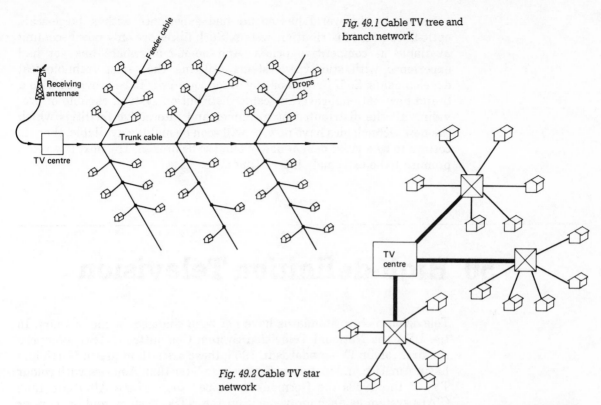

Fig. 49.1 Cable TV tree and branch network

Feeder cable

Receiving antennae

Drops

Trunk cable

TV centre

TV centre

Fig. 49.2 Cable TV star network

possibly because it is comparatively expensive. An interactive system provided mainly for entertainment purposes is perhaps unlikely to suffer such a slow rate of growth.

Basically two different types of cable TV network can now be considered. All existing networks are designed on the "tree-and-branch" principle; but thanks to advances in technology the "star" principle is now a possible alternative.

1) *Tree and Branch* (fig. 49.1)

All programmes are initiated at or fed into the system from a single main centre; all programmes are fed out into the network from this centre. Wide-bandwidth cables are needed into every house. Selection of programmes to watch is carried out at the TV receiver itself since all the programmes are carried into every house.

2) *Star* (fig. 49.2)

All programmes are initiated at or fed into the system from a single main centre, and all of them are fed from this centre to a number of "intelligent" distribution centres. Wide-bandwidth cables are needed for these links. Cables between these distribution points and users houses need not have such a wide bandwidth as the main links because only the particular programmes actually requested by the user, by keying code buttons on the TV set itself, will be sent out on the local line to the user. Conventional cables would therefore be adequate for these local links. This star-type distribution system is somewhat more flexible than a tree-and-branch system in its ability to provide for new features—particularly interactive features.

No-one in the world has so far had experience with a large-scale optic-fibre-based distribution system. Such fibres are only now becoming available at competitive prices. And no-one anywhere has yet had experience with star-type systems utilizing the latest technological developments. So if Britain or any other country is to be provided with a brand new national-coverage cable distribution system, capable of providing all the distributed programmes and interactive facilities which the new technologies have now, or will soon be making available, there is certain to be a great deal of new ground to be broken. The next few years promise to be extremely interesting ones.

50 High-definition Television

Television picture standards have not been changed for many years. In the USA the National Television System Committee (NTSC) produced the first colour TV standards in 1953; these are still in use in North and Latin America and in Japan. Europe was later than America with colour TV; in the 1960s the Germans developed their Phase Alternate Line (PAL) system as an improvement on the NTSC system, and at more or less the same time the French developed their Sequential Couleur a Memoire (SECAM) system. The two systems are in wide use in Europe, Africa and Asia. These two European-developed colour systems offer considerably better resolution and colour quality than the original NTSC system; ten years made a tremendous difference.

Twenty more years have now passed. If a colour TV system were to be designed and manufactured today without reference to the standards of the 1950s and 1960s, it would undoubtedly be a very different system from NTSC, PAL and SECAM, and would provide picture quality as good as that of the best colour movies.

The use of HDTV practices is in fact likely to become common in the moving picture industry before HDTV really hits the broadcasting world. The reasons for this are that electronic editing at computerized consoles is much cheaper and quicker than having to develop film and cut it. Titles and special effects can be added electronically far more cheaply than by using "traditional" procedures. The use of HDTV technology in the movie industry is, however, in effect a closed-circuit use within the industry; it does not affect TV broadcasting stations, bandwidth economy or home TV receivers.

The NTSC system needs a bandwidth of 6 MHz; PAL and SECAM both need 8 MHz bands. For comparison, an HDTV system recently demonstrated in Japan has a video signal bandwidth of 27 MHz and uses a radio bandwidth when transmitted by FM radio of 100 MHz. This system gives pictures with a 1125 line structure (cf. NTSC 525 lines, PAL/SECAM 625 lines) which are reported as being equal in quality to 35 mm colour film and could be projected on to wide screens (2 metre diagonal) with complete clarity and sharpness of image. This NHK (Japan Broadcasting Corporation) HDTV system is already however slightly "old hat" in that it uses analog signals throughout; many designers are now looking into the economics of adopting digital techniques for both HDTV video and audio services.

The use of HDTV display techniques by national ISDN systems is also under close consideration. When optic-fibre local-distribution networks become available, the higher bandwidth needed for HDTV will present no major difficulties: it will then become possible for top-grade colour reproduction on big screens to be available in any home.

Since TV picture quality in NTSC countries is worse than that in PAL/SECAM countries, and since the electronic industries in the USA and Japan have traditions of being able to take extremely rapid action, there could well be more pressure to introduce HDTV in Japan and the USA than in Europe. All that can safely be said at this time is that those who are first into large-scale production of HDTV will doubtless set standards which the rest of the world, coming along later, will either grudgingly be forced to follow or will ignore, hoping that the more time that elapses the greater the chance that it will be possible to agree on a single standard to be used by all countries of the world.

51 Credit Cards and Smart Cards

Metal coins and paper currency may be a bit of a nuisance (although a wallet full of crisp new high-value notes would no doubt be very acceptable to most of us) but in the more developed countries of the world there is a well-established move away from the use of hard cash. Major payments have of course for many years been carried out using cheques or bank drafts, but there is now also a steady increase in the use of "plastic money", credit cards, for relatively small payments.

A high proportion of hotel, airline and restaurant bills in internationally popular tourist areas is now paid by credit cards and the use of such cards for ordinary retail purchases is now becoming generally acceptable in many countries: so much so that in some areas of the USA a man who pays cash for goods is somehow rather looked down upon, he is thought to be someone who must have been turned down by credit card organizations as not being credit-worthy, not a good-risk!

When a purchase is made using an ordinary credit card (American Express, Diners, Visa, Master Card, etc.), the sales assistant sometimes vanishes for a short while; she has had to make a telephone call to the office of the credit card company to make sure that the particular charge concerned may be accepted. Most ordinary credit cards do however usually have a backing strip of magnetic material. When the card is placed in a special reader, the card number can be read out, automatically. In many countries, special telephone instruments are now available. The credit card is wiped along a slot in the back of the instrument; this automatically sets up a call to the credit card company concerned and obtains the necessary authorization without the shop assistant having to carry on a whispered—and possibly embarrassing—conversation.

When Integrated Services Digital Networks (ISDNs) have been brought into service, it is possible that services such as authorizations of credit card sales will be carried out on a packet switched basis. The small amount of data to be transmitted and received does not need the establishment of a full speech path, the 56 kbit/s or 64 kbit/s of a normal telephone call set-up; data at 8 kbit/s is perfectly adequate for these "Point of Sale" (POS) transactions. When ISDNs have become widely available, POS instruments able to read credit cards, send coded information straight to the card company computer, and receive its reply, all automatically, should overcome many of the difficulties now experienced with credit cards both by consumers and by retailers. The huge lists of lost, stolen and invalid cards, almost as big as some telephone directories, will for example no longer have to be ploughed through before a $10 purchase can be agreed.

But all these transactions still use paper as their final authority, a signature is still needed on the credit card payment slip.

Some telecommunications administrations, faced with huge losses by the vandalism of public payphones (coin box telephones), have recently introduced semi-intelligent cards which can be purchased for cash and which can be inserted into a slot in a payphone to permit calls of up to a predetermined total cost to be established. The balance still available after a call has been completed can be read off the card by the payphone into which the card is next inserted, and so on until the whole amount has been used up. The card is then thrown away.

It is only a small step from these payphone cards to what are now being called Smart Cards. These look like ordinary credit cards but they have a built-in electronic memory and the logic circuits of a microprocessor controlling read and write access to this memory.

Trials are under way of several different types of Smart Card; there is as yet no universally acceptable standard. For use in ordinary retail transactions, the Smart Card acts like an Electronic Cheque Book. The user has to remember (and key in) a Personal Identification Number (or PIN). The card's memory is programmed to know the total value of all purchases that may be made during each month. It authorizes and records all transactions up to this limit, and by placing the card in a special electronic reader, available at Point of Sale, the holder can see a display of current outstanding credit. This pre-set "revolving credit" limit is restored monthly. When a purchase is made, the POS terminal in the shop records all relevant information. Then when the day's work is over or at any other convenient time, all the data is transferred automatically to the bank for crediting the shop's account and debiting the customer's account. It can be seen that this will not only cut administration costs but will also make Smart Card sales much more acceptable to small money-hungry shopkeepers than credit card sales are now. In some countries it can take three weeks or even more before payment is actually received by the shopkeeper from the credit card company, so discounts which are freely available for cash purchasers are sometimes not given to credit card users. This naturally holds credit card sales down.

By the end of this 20th century many of the big shops in most developed countries could well therefore be operating on a completely non-cash basis. There would be no need for citizens of rich developed countries to carry pockets full of heavy money. A reserve supply of a few small coins (to permit casual purchases of candy bars and magazines) is all that will be needed to back up their Smart Cards. For many people in third world countries of course (and for many in rich countries also), even a pocket half full of money will, regrettably, still be beyond their wildest dreams but this will not be the fault of the telecommunications engineer!

52 Direct Broadcasting by Satellite

Early telecommunications satellites had very little power: in order to pick up their signals (remember that these have to travel nearly 36 000 kilometres to reach the earth), very large diameter high-gain dish antennae were needed. Ground stations working to Intelsat satellites still need dishes at least 11 metres in diameter.

With improvements in technology it is now possible for satellites not only to put more power into their transmitted signals but also to feed this power into specially designed antennae on the satellites themselves, which focus all the power on to a comparatively small part of the earth's surface. (Early satellites had to accept that their transmissions were directed towards the whole face of the earth.) This ability to focus radio power on to a small ground area means that a receiver on the ground no longer requires a huge dish antenna. A very small unit only a few centimetres across is perfectly adequate to give a good signal/noise ratio and an acceptable TV picture. These small antennae are of course for receive-only stations; it is not at present possible to transmit back up to a satellite using tiny dishes.

The use of this new generation of satellites (with higher power, greater bandwidth and special directive antennae) has therefore now made it possible for satellites to be used to broadcast programmes direct to users. Every home may well soon have its own small dish looking up at a geostationary satellite poised over the equator, beaming TV programmes down to continents and countries. European countries are to be served by satellites stationed over the South Atlantic (see fig. 52.1). National policies

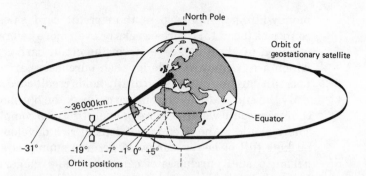

Fig. 52.1 Broadcasting satellites on geostationary orbit (*Courtesy: Telcom Report 5, 1982, Siemens*)

may well ensure that coverage zones for all the different services will approximate closely to national boundaries (see fig. 52.2).

Each country in Europe has now been allocated channels in the 12 GHz band to enable Direct Broadcasting from Satellites (DBS) to be introduced. It seems likely that DBS services will begin to become available by 1986. Each channel will be able to carry a TV signal and two or more accompanying sound signals—either stereo sound or two different sound programmes, in different languages.

There is however one major difficulty. No TV standard in current use takes account of the tremendous advances which have been made in technology in recent years. As already explained, the NTSC colour system used in America and Japan was, for example, invented in 1953; the PAL system used in Britain and Germany was first used in 1963, and the French SECAM system at about the same time. All these colour encoding systems require luminance and chrominance signals to share part of the frequency band. This latter feature frequently results in colour reproduction which is not really as good as it should be. Flashes of false colour on fine patterns (cross-colour) and spurious dot patterns on moving edges (cross-luminance) are often experienced.

DBS is going to represent such a huge capital investment that it would clearly be desirable to start off with systems which recognize the latest technological developments. Since satellites by their very nature are not restricted to providing coverage within a national boundary it would also be an attractive idea if all the world's TV authorities could agree on a single new system to replace all three rival systems, NTSC, PAL and SECAM.

Here there is another difficulty, one of timing. DBS services are becoming available too soon for there to be any realistic chance of reaching international agreement on so fundamental a series of changes, in time for DBS to start off with a brand new one-world system. The French, for example, are probably going to start DBS off in 1986 using their SECAM, and the Germans are likely to continue to use PAL. Just as when colour TV was first introduced it still had to be possible for older black-and-white TV sets to continue to be used, there is great pressure to ensure that any improvements made possible by DBS have no adverse effects on users of currently available TV sets.

In order to eliminate undesirable cross-colour and cross-luminance effects, various proposals have been made. Britain's IBA have put forward a proposal (called MAC, Multiplexed Analogue Component) to

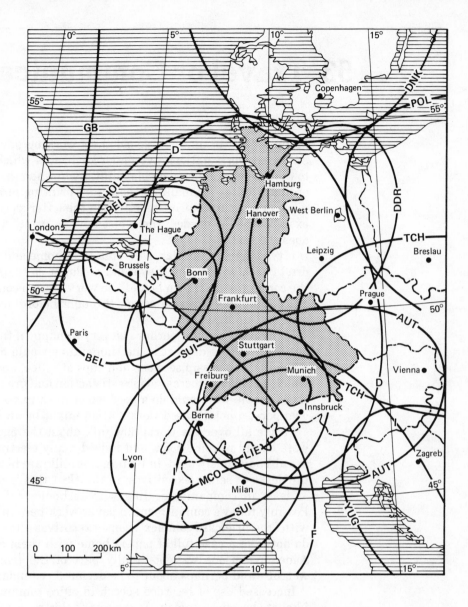

Fig. 52.2 The coverage areas for the satellites of various European countries (*Courtesy: Telcom Report 5, 1982, Siemens*)

separate luminance and chrominance in time, i.e. to broadcast these signals on a time-multiplexed basis. The BBC has proposed an Extended PAL System which separates these signals by putting them into separate frequency bands, using the wider frequency bands that DBS will make available.

Whatever system is ultimately adopted will have to take into account the possibility of future developments—by the mid-1990s it may well be economically possible for extremely high-quality TV to be provided, possibly using digital technology. Broadcast TV (with its need for great economy in the use of the available frequency spectrum) is also bound to be affected very greatly by the growth of cable TV (in which frequency economy may not always be so pressing a requirement).

53 Travel or Communicate?

Many office workers now spend their working days sitting before computerized work-stations, typing on keyboards which print characters on to the screen of a visual display unit. Even when the task is to prepare an ordinary letter for mailing, the actual printing process now often takes place in a separate print-room; when the letter as shown on the screen is exactly what is needed, the typist keys a "print" key and the computer does the rest.

If typists had a similar work-station at home and took a batch of rough drafts back with them, they could no doubt do their work just as well in the comfort of their own home. The work-station could be linked with the office computer when necessary through the ordinary telephone system, on a dial-up basis.

Many people might perhaps be very unhappy if they had to spend most of their working days at home, alone: they would miss the social side of office life, the gossip sessions and cups of coffee, and all the interactions which can lead to increased job satisfaction and greater efficiency. But for those people who could do a good job of work in their own time, at their own pace, a home-based work-station might be an ideal arrangement.

People all over the world now talk about the paper-less office as if it were bound to come soon, and indeed many electronic business devices are already widely used. In Britain it is still rare however to see VDUs on the desks of senior executives; in the USA this is not at all uncommon, at least in companies dealing with technologically advanced matters. Possibly today's commitment to paper-work here in Britain is associated with "the briefcase complex"; some executives appear to feel that, if they do not take a case full of papers home with them for study, they are in some mysterious way not really part of the team, not pulling their weight—and perhaps about to be declared redundant.

Increased use of recorded speech in office communications is however likely to become common. As the cost of electronic memories falls and the cost of human secretarial services rises, it would seem logical to be able to store messages digitally in the form of speech. These messages could go straight to the addressee as soon as he or she becomes available and thus avoid processing by secretaries. They would of course still remain in store for reference as long as necessary. The humble telephone answering machine is a small step in this direction. Some executives still "dry up" and promise to call later rather than leave a message on an answering machine, but familiarity with such devices rapidly results in user-confidence.

It is however in reducing travelling time and costs that electronics really comes into its own. Many companies now use telephone conference calls involving several executives, at different locations. A third party can be brought into a conversation merely to clear up a point under discussion. Most modern telephone switching systems provide features of this type.

If meetings are held regularly involving staff from two different locations, it is often economic to equip rooms specially as conference studios using a form of closed-circuit television called Confravision or Teleconferencing to link the two locations. Cameras can if necessary be directed to focus on the person doing the talking or can be used to give close-ups of plans or drawings which are the subject of discussion. It takes very few such conferences to save the money otherwise spent in travelling and hotel accommodation.

Radio-paging is already common in many countries. Some of these will merely "bleep" to indicate that the user should phone the office; others can give simple coded messages on a strip of LCD (liquid crystal display, like a digital wrist watch) or can even pass on a voice message as soon as the user indicates that this can safely and securely be done.

The use of viewdata services such as Prestel is another efficient way of eliminating the need for physical travel. Prestel can be used on a completely interactive basis to enable questions to be put to people at a distance and their answers obtained immediately. It is highly likely that, within a few years, enterprising merchants will establish Prestel calls to major customers every morning so that their orders for goods may be placed for rapid delivery, with no paper copies of orders at all, no travelling to warehouses, and direct charging to the purchaser's account with no bills and no cheques.

54 Corporate Communications

Presidents, vice-presidents, directors and managers of corporations and public companies are nowadays exposed to a great deal of media comment about telecommunications. Telematics—the "marrying together" of telecommunications and computer technology—is regularly the subject of searching and informative articles in business magazines and management journals.

How can companies best take advantage of all the new-technology opportunities which are now beginning to become available? Is the "no-paper" office already a practicable proposition, with each manager sitting before a flickering screen and a row of keys? An honest answer has to be: "No, completely paperless offices are not really economic for us all, yet."

Some trades and professions do however lend themselves to more rapid updating than others. Most tasks which are primarily concerned with figures and with interpreting them can, even in today's "interim world", be carried out with fewer personnel if the power of readily-available computers is harnessed to perform those tasks for which they are specially suited. A personal computer programmed to help solve the problem which a particular manager has to tackle is clearly one step forward in the direction of full automation.

Many companies now employ computer managers or data managers whose duties include studying procedures and practices in all departments and bringing automation and computers into use—either centralized on or a distributed basis—wherever overall efficiency could by such means be improved.

A very real problem is, however, getting information from one place to another, the telecommunications side of telematics, and having got it there to ensure that it is in a form which is understandable to the recipients so that they may study all relevant factors, reach all appropriate decisions, and take all necessary action.

Until recently only the largest of companies had Telecommunication Managers of their own: telecommunication was in most companies merely a matter of liaising at irregular intervals with the local telephone company and the local telex organization and renting from them the equipment which they said was appropriate to what they saw as the company's needs.

Beginning in the 1970s and now in the 1980s business conditions all over the world have become more competitive—managers can take nothing for granted any more and are forced to look into details of all expenditure. Sometimes they found for example that the interests of their local telephone company were not always exactly the same as their own corporate interests. Even though the local TelCo was a monopoly supplier, its tariff structure was often extremely complex so it became sensible for major companies to lure specialists to help them get value for money.

These early Telephone Managers were often men just retired from the local TelCo, people who were completely familiar with the local telephone tariff structure and could not only work out the most cost-effective packages for their new employers but could also—through the "old boy network"—ensure that any work which was needed was done speedily.

In America, de-regulation and interconnection were really the political triggers which made genuine competition become possible. Telephone Companies very reasonably fought to retain as much of their old monopolies as possible but long-distance routes were the first to become truly competitive. Companies like MCI were (often after much expensive litigation) able to take advantage of the fact that the whole US national telephone system, the biggest and most complex in the world, had been built up as a network of local, intra-State, inter-State and overseas operating organizations, with cross-subsidization of local telephone companies from the considerable revenues earned by long-distance facilities. Specialized common carriers such as MCI, Sprint and many others are now able to offer users a choice of routing for long-distance calls, often at considerably less cost than the standard TelCo Direct Distance Dialling (DDD) charge per call. Telephone companies have naturally fought back by introducing many attractive new tariff packages, most of which enable major companies to reduce their telephone bills but have little effect on ordinary domestic users and small businesses. The Wide Area Telephone Service (WATS) is perhaps the best known of these: by paying a flat-rate monthly fee (considerably higher than that for an ordinary line), long-distance calls within certain geographic zones and up to a specific total number of minutes per month are all uncharged. The average charge per minute (provided your monthly allocation is fully used) is much less than if long-distance calls had been dialled out as ordinary DDD calls. The task of the Corporate Telephone Manager therefore became more complex: time was no longer spent arranging for key systems and extension instruments to be moved from one office to another but in working out the most economic mix of long-distance circuits for each location in which the company had an office or plant: so many WATS lines, so many to MCI, right down to the last choice bundle, so many ordinary DDD lines. Fortunately the advent of the microprocessor has now put a management tool of great efficiency into the Telephone Manager's hands. Many modern PABXs can (when correctly programmed) operate on a "least-cost routing basis"—they choose whichever circuit is the cheapest at that particular time, and companies such as TDX are able to offer least-cost-routing packages to smaller users whose own installations are unable to provide such choice of routing facilities. Our Telephone Manager does therefore now have to become something of a software expert; programming has suddenly become an essential part of duties.

In Europe, developments are proceeding rather more slowly than in the US and with a slightly different emphasis. In Britain, for example, the national telecom system has for many years provided long-distance circuits for customers on a relatively cheap monthly rental basis so that company PABXs in different parts of the country are often interconnected with a complex mesh of private trunk circuits. This means that the majority of inter-branch and inter-city calls do not need to switch out into the Public Switched Telephone Network (PSTN) at all. There is therefore no real demand for least-cost routing equipment of the type which has recently been developed in the USA, although management information services which are often a side-product in least-cost routing systems are proving extremely popular in Britain in that they enable individual telephone extensions to be debited with the cost of each long-distance call made.

Political action in most European countries seems to be leading slowly to the breaking up of the traditional Government-owned telecommunications monopolies and their conversion into public limited companies; this changeover is strangely enough usually called privatization, because the end result is the transfer of financial responsibility from the public sector to the private sector. In Britain, privatization has gone one stage further than in other European countries: not only has British Telecom within a few years been changed from Government Department (part of the Post Office) to quasi-Government Board to Public Limited Company, but a completely separate national telecommunications company, the privately owned Mercury Communications Ltd, has been founded, and licensed by the Government to provide services, so that there will ultimately be effective competition, particularly for the nation's business communications, between Mercury and British Telecom. BT is now Britain's largest company, with almost 250 000 employees, whereas the new rival company at last count had a total of about 300 employees (only 50 in 1983), so it is, at least initially, something of a David-versus-Goliath situation. But Goliath has a large millstone round his neck: a basically analog national system, many hundreds of small rural exchanges and their associated far-flung local distribution networks whereas Mercury does not have to spread out into the countryside and is able to start up with a completely new technology network (digital microwave, optic fibres, digital switching).

Having become something of a software expert when Stored Program Control (SPC) came in, in the 1970s, our corporate Telephone Manager is now beginning to take responsibility not just for corporate telephones but for all the company's telecommunications: for the devices which talk electrically to each other, for the computers, the terminals, the communicating word processors, and, in effect, for the whole office automation field. What started off as an easy sinecure of a job to be filled by a pensioner now demands a completely professional *Telecommunications Manager*.

Wang laboratories, one of the foremost organizations in this business has defined Office Automation as being the development and application of six technologies:

a) *Data Processing:* information in the form of *numbers.*
b) *Word Processing:* information in the form of *written words.*
c) *Image Processing:* information in the form of *pictures.*
d) *Audio Processing:* information in the form of *spoken words.*
e) *Networking:* the way these various forms of information are communicated from one local to another.
f) *Human Factor:* the end result of office automation is to make people more productive through the use of technology. Technology for its

own sake is no use. It must be applied in a way which is acceptable and genuinely helpful to people.

All six of these are closely involved in nearly every major project with which Telecoms Managers have to deal and every project has to be planned so that machines do what machines do best and people do what people do best.

Two examples of corporate communications work, one from the USA and one from Europe, will indicate the variety of tasks which now have to be handled by Telecommunications Managers.

The European example chosen is the telephone network operated by a major UK oil company: it, with its subsidiaries, has major installations in more than 90 towns, in all parts of Britain. Most of the local telephone systems and data terminals at these locations are at present linked together using rented voice-grade circuits. Traffic patterns in the whole network are being studied and forecasts of future traffic prepared so that an economically-efficient fully-integrated corporate network may be planned, taking advantage of the latest available digital technologies.

The American example is the private telephone system of the University of California, Los Angeles. UCLA has in effect just become its own telephone company. Prior to 1983 all the phones in UCLA's 411-acre campus were provided, installed and maintained by the local telephone company, General Telephone of California, with switching performed by a relatively old step-by-step office. Now, UCLA has its own digital exchange, serving some 13 000 instruments in 82 different building complexes all over the campus. New technology (a Northern Telecom SL-100 switch and a DEC VAX 11/750 computer system) provide detailed management information including itemized billing to all the many cost centres in the university. Each cost centre (1201 of them) is now fully accountable for all the costs it incurs. Operators have immediate access to up-to-date directory information—mobility is high in California and changes are frequent. UCLA has its own Police Department and the new exchange provides its own 911 emergency call system with immediate display on a VDU at the Police Station of the location of any instrument initiating an emergency call. This new system has cost the University nearly $20 million but when all the running and maintenance costs are taken into account they expect a net saving of $15 million over the next 15 years.

More and more national telecommunications systems will soon be providing an Integrated Services Digital Network (ISDN) by which a 56 kbit/s or 64 kbit/s path can be established right across a country from one office to another, useable either for switched voice services (the telephone) or for switched data services (including teletex, packet switching, etc.). More and more manufacturers are already able to offer Local Area Networks (LANs) to connect together all the digital devices in a building complex: voice, data, facsimile, computers, VDU, remote terminals, etc.

The task of the Telecommunications Manager is to plan and operate these exciting new networks so that overall corporate efficiency will be maximized.

Index

acceptor element   129
A/D conversion   101
address   127
ADI   108, 109
aerials   5, 68
Agana   60
air pressurization   52
A-law coding   102
alternating current   12
alternator   12
ALU   142, 143
aluminium   52
AMA   85
American Express   155
AMI   108, 109
amplifier   1
amplitude modulation   24
analog   46, 97
analog versus digital   97
AND gate   130
angular rotation   15
antennae   5, 68
anti-coincidence gate   131
ANZCAN   60
apportionment of reference
   equivalents   52
arithmetic growth   87
armouring   57, 58, 59
artificial voices   148
associated signalling   42
Atlantic Ocean Region   73
attenuation   5, 21, 66
Auckland   60
augmentation   118
automatic gain control   66
automatic telephone exchanges   9, 31,
   35, 39, 46

back porch   92
back-up store   134
Baler   60
band-pass filters   18
band-stop filters   18
bandwidth   20, 24, 25, 26, 55, 61, 64
bank   31, 32
balanced modulator   24, 55
baseband   120
BCD   98
Beaver Harbour   59

Bel   11
bi-directional   2, 50
billing   84
binary   97–101
binary coded decimal   98
bi-quinary   101
bit   98–101, 127
bit rate   64
blanking period   92
blocking   38
blue gun   96
Boolean algebra   130
BORSCHT functions   49, 50
BRACAN   59
braided ring   122, 123
broadband   120
bubble   134
buffer   122
bulk billing   84
burden of spare plant   54, 89
byte   127

cabinets   54
cable television   120, 152
Cairns   60
call accounting   81
calling channel   8
CAMA   85
Camuri   59
Canary Is.   59
CANTAT   59
capacitance   4
capital costs   83
car phones   78
carrier   2, 24, 25, 55
Cassegrain antenna   71
catenary   52
CATV   152
CCIR   150
CCITT   30, 42, 77, 145, 149, 150
Ceefax   149
cells   79, 80
cellular radio   78
centralised maintenance   81
Centrex   45
CEPT   150
channel   1, 2, 55, 61, 106
channel filters   55
chrominance   93

Chinen   60
circuit   2
circuit switching   111, 113
civil works   86
cladding   61–63
Clarenville   58
CMOS   132
coaxial cable   2, 4, 5, 58, 59
cochlea   19
codec   49, 107
codes, magnetization   137–38
coherent light   60
coincidence gate   132
collision detection   121
co-located concentrator   52
colour TV   93
COLUMBUS   59
commercial speech   17, 20
common channel signalling   42, 43,
   50, 119, 145
common control exchanges   36, 39
communicating word processor   112
compatible signal   93
compelled signalling   41
complementary colours   94
COMPAC   60
compound interest growth   87
computer bureau   110
computers   39, 40, 109, 141
concentration stage   47, 48, 52–54
conductors   2, 3, 4
Confravision   161
congestion   38
Conil   58
contention   121
copper centre   86
core   61–63
core memory   134
Corner Brook   59
cosmic noise   23
cost comparisons   87–89
CPU   142
credit cards   155
crossbar switches   36
crosstalk   23
CRT   90, 95
CSMA   121
CVSDM   30
CWP   112

cycle 13
cyclic redundancy check 139

datagram 115
data services 109
DBS 157
DCE 116
DDD 163
debugging 82
decibel 10, 11
decoding 93, 107
delta modulation 30
demand forecasting 85
demodulator 56
deviation 27
diagnostics 81
dial 31
dielectric 4
dielectric losses 21
diffusion 129
digital exchanges 46
digital fundamentals 127
digital islands 118
digital signals 97
digital strategies 118
digital techniques 97
DIL 128
Diners Club 155
DIP 128
dipole antenna 68, 69
direct broadcasting by satellite 157
direct current 12
directors 69
discrete components 128, 132
disc stores 134, 137
dish antenna 6, 70
dissociated signalling 42
distortion 1, 21, 105, 109
distress calls 76
distributed control 39
distribution cables 51
distribution networks 51
D layer 65–67
donor element 129
DTE 116
DTMF 41
dual ring 122, 123

EAPROM 136
EBU 150
ECL 132
E layer 65, 66
electromagnetic wave 5
electro-mechanical switch 32
electronic cheque book 156
electronic common control 39
electronic mail 144
empty slot 121
encoding 93, 107
envelope 25

EPROM 136
error checksum 115
etching 129
exchanges 8, 9, 31
expansion stage 49
exponential growth 87
external plant 51

fading 66, 67
facsimile 125
fan-in 133
fan-out 133
FDM 24, 25, 47, 55, 145
field, magnetic 4
filters 18
financing costs 83
first-choice selectors 34
first line maintenance 82
flat rate 83
F layer 65, 66
flexibility points 51
flip-flops 135
flicker 90
floppy discs 140
fluctuation noise 23
forecasting 51, 85
Fourier 16
frame alignment 107
frequency 12, 24–28, 69, 92, 95, 103
frequency changing 79
frequency deviation 27, 28
frequency division multiplexing 24, 25, 46, 145
frequency modulation 27, 92
frequency ranges 17, 18
frequency spectrum 25, 56, 92, 95
front porch 92
fundamental frequency 20

galactic noise 23
gates 104, 105, 130
geostationary satellites 72
Gigahertz 16
graded index 62
green gun 96
Green Hill 58
ground wave 67
group 55, 56
Guam 60
guns 96

Hanauma Bay 59
harmonics 17, 19
HAW cables 59
Hawaii 59
HDB3 108, 109
HDLC 116, 117
HDTV 154
heat generation 12
heat losses 21

hertz 14
hexadecimal 99, 100
high-definition tv 154
higher-level multiplexing 106
high pass filter 18
high tensile steel 58, 59
Home Office 160
Hong Kong 60
horn antenna 71
human hearing 17, 19, 20
hybrid 49, 107
hypergroup 57

IBM 129
ICs 128
in-band signalling 41
Indian Ocean Region 74, 77
inductance 4
information technology 126
infra-red 18
insulation 3
insulators 3
Inmarsat 76
Intelsat 72, 157
interactive cable tv 152
interactive data 114
interference 1, 64, 97
integrated circuits 128
Integrated Services Digital
  Network 124, 165
intersticial quads and pairs 5
inverted channels 56
ionisation 65
ionosphere 6, 7, 64–67
I/O devices 142, 143
ISDN 43, 84, 124, 156, 165
islands (digital) 118
ISO 116
IT 126
itemised billing 85

jelly-filled cable 52
Johnson noise 22
joint use of plant 52
junctions 8

Keawaula Bay 60
Kelvin temperature 22
keyphones 44, 45
kilohertz 16
Kota Kinabalu 60

LANs 110, 119
laser 60
latches 135
layered protocol 116, 117
layers 65
LCDs 161
leased circuits 110
LEDs 60

lightweight cable 58, 59
Lincompex 67
lines 2, 51
local area networks 110, 119
local distribution cables 51
local line networks 51
log-periodic antennae 70, 71
logic gates 130
low pass filters 18
LSI 129
luminance 93, 96

MAC 158
magnetic bubbles 134
magnetic field 4
magnetic tapes & discs 134, 137
magnetization codes 137, 138
main cables 51–53
maintenance of exchanges 81
mainframe computers 141
main stores 134
mail services 145
Makaha 59, 60
make-and-break 31, 40
MANs 119, 120
maritime communications 76
marker 36
master card 155
master group 57
matrix 37, 38, 134
MCI 163
measured rate 83
Megahertz 16
memory 127, 134
message accounting 83
message rate 83
message switching 111
metering 84
MF signalling 41
microcomputers 141
microprocessor 109, 141
microwave links 2
Midway 60
minicomputer 141
mobile radio systems 78
modem 110, 151
modulation 24
modulator 55
MOJ (metering over junction) 84
monochrome 93
monolithic ICs 129
monomode 61–63
MOSFET 132
MSI 129
mu-law coding 102
multi-frequency signalling 41
multi-hop transmission 67
multimode 61–63
multimetering 84
multiplexing 2, 55, 104

music 21

NAND gate 131
network management 81
nibble 128
Ninomiya 60
noise 21–23, 97
noise margin 133
non-intelligent terminals 114
non-volatile memory 135
NOR gate 131
Norfolk Is. 60
NOT gate 131
NRZI code 138
NTSC tv 93, 154, 158
numbering plan 117
numbering systems 98

Oban 58, 59
octal 99, 100
off-hook 40, 49
office automation 164
on-hook 49
operating costs 83
optic fibres 2, 60
Oracle 149
OR gate 131
outband signalling 41
operation and maintenance
    services 81
overlay principles 118, 119

PABXs 44
Pacific Ocean Region 75
packet switching 113
PAD 114
paging 161
pairs (of wires) 2
PAL 93, 154, 158
PAM 29, 102
parabolic antennae 68, 70, 71
parasitic elements 69
parity check 139
partition noise 23
PAX 43
PBX 43
PCM 30, 46, 48, 49, 101
Penmarch 58
periodic pulse metering 84
periodic time 14
persistence of vision 90
phase 15, 17
phosphors 96
Picture Prestel 150
pitch 20
planar transistor 129
planning 86
PMBX 44
Point Arena 59
point of sale 156

power of speech 19
power ratios 10
Prestel 149, 161
primary cables 51–53
primary colours 94
primary multiplexing 106
pressurisation of cables 52
PROM 136
propagation delay 133
propagation, radio 64
protocol 116
pseudo-digital 110
PSTN 110, 111
PTH 129
public switched telephone
    network 110, 111
pulse code modulation 30, 46, 48, 49,
    101
pulse modulation 29
push button dialling 41

quad cables 3
quantizing 30, 101–103
quantizing noise 102, 104
quasi-associated signalling 42

radio antennae/aerials 68
radio propagation 64
radio systems 5
RAM 135
random access 135
read head 137
Recife 59
recombination 65
red gun 96
redundancy 82
reed relays 35–37
reflectors 69
Reeves, Dr 101
refractive index 62, 63
regeneration 105–107
register 36, 134
register insertion 122–23
relays 9, 34–38
reload (software) 82
remote concentrator 52–54
repair services 82
repeaters 57
repeater section 64
repeat multi-metering 84
replacement policy 118
resistance 4
resistor noise 22
reverse-channel capability 152
rhombic antenna 68, 70
rings (data) 110, 121–23
rollback (software) 82
ROM 135
rotary dial 31

sampling   101–103
San Luis Obispo   59
satellite broadcasting   157
satellites   5, 70, 72, 76
saturation   94
scatter   5
Schottky   133
SCOTICE   59
SEACOM   60
SECAM   93, 154, 158
secondary cables   51–53
selective fading   67
selectors   32–34
semi-logarithmic scales   88
service management   81
ships radio   76
shot noise   22
sidebands   25, 55
side frequency   25
signalling   40
signal transfer point   42
signal/noise ratio   23, 80
silica   61
Singapore   60
single sideband   26
sinusoidal waveform   13
SITA   109, 111
skin effect   4
sky wave   5, 66, 67
smart cards   155
SMDR   45
software   82
space division   46
space wave   5
SPC   39, 164
speech   19
square wave   17
SSI   129
star networks   122–24, 153
static   23
STD   84
step-by-step switching   31
stepped index   62
St Hilaire de Riez   58
store-and-forward   113, 114
stored program control   39
stores   134
straight line growth   87

Strowger   34
submarine cable   57
subscribers' line interface   49
subsystems   47, 48
supermaster groups   57
supergroups   55
surface wave   5
Suva   60
SWIFT   145, 146
switching   7, 31
switching, data   111
Sydney   60
Sydney Mines   58
synchronisation   91, 96, 105

tariffs   83
TAT   58
TCM   129
TDM   46, 104, 145
TDX   163
telecommunications   1
Telecommunication Manager   162
teleconferencing   161
telematics   126, 162
telephone tariffs   83
teletex   112
teletext   149
television   90
telex   112
Telidon   150
thermal noise   22
threshold of feeling   19
threshold of hearing   19
time division   46
time division multiplexing   104
token system   122, 123
toll ticketing   85
torn paper   111
traffic list   76
transducer   1
transformer   4
TRANSPAC   60
travel or communicate   160
troposcatter   6, 7
tree-and-branch network   153
truth tables   130–32
TTL   132
Tuckerton   58

TV   90
two-motion selector/switch   32
two-wire line   2

UCLA   165
ultra-violet   64–65
unidirectional   2
uniselector   32
unit twin cable   3
usage sensitive charging   83

VDU   112, 142, 143, 160
velocity   16
vestigial sideband   92
VF signalling   41
viability   83
videoconferencing   161
Viewdata   77, 149
video signal   80
videotex   125, 150
violation mark   108
virtual call   115
Visa   155
visible light   18
voice signal   90
VLSI   129
VMS   148
voice frequencies   17
voice frequency signalling   41
voice messaging   147
voice recognition   148
volatile memory   135

WANs   119
water barrier   52
WATS   163
waveform   16
wavelength   16
Widemouth Bay   58, 59
wiper   31–33
word   127
word processor   112
write head   137

X rays   18
X25   116, 117

Yagi antenna   68, 69